Alternative Energy

Alternative Energy

Volume 1

Neil Schlager and Jayne Weisblatt, editors

U·X·L

An imprint of Thomson Gale, a part of The Thomson Corporation

Detroit • New York • San Francisco • San Diego • New Haven, Conn. • Waterville, Maine • London • Munich

Alternative Energy

Neil Schlager and Jayne Weisblatt, Editors

Project Editor
Madeline S. Harris

Editorial
Luann Brennan, Marc Faeber, Kristine Krapp, Elizabeth Manar, Kim McGrath, Paul Lewon, Rebecca Parks, Heather Price, Lemma Shomali

Indexing Services
Factiva, a Dow Jones & Reuters Company

Rights and Acquisitions
Margaret Abendroth, Timothy Sisler

Imaging and Multimedia
Randy Bassett, Lezlie Light, Michael Logusz, Christine O'Bryan, Denay Wilding

Product Design
Jennifer Wahi

Composition
Evi Seoud, Mary Beth Trimper

Manufacturing
Wendy Blurton, Dorothy Maki

For permission to use material from this product, submit your request via Web at http://www.gale-edit.com/ permissions, or you may download our Permissions Request form and submit your request by fax or mail to:

Permissions
Thomson Gale
27500 Drake Rd.
Farmington Hills, MI 48331-3535
Permissions Hotline:
248-699-8006 or 800-877-4253, ext. 8006
Fax: 248-699-8074 or 800-762-4058

Cover photograph for Volume 1 Oil drilling platforms off the coast of Texas © Jay Dickman/Corbis.

While every effort has been made to ensure the reliability of the information presented in this publication, Thomson Gale does not guarantee the accuracy of the data contained herein. Thomson Gale accepts no payment for listing; and inclusion in the publication of any organization, agency, institution, publication, service, or individual does not imply endorsement of the editors or publisher. Errors brought to the attention of the publisher and verified to the satisfaction of the publisher will be corrected in future editions.

LIBRARY OF CONGRESS CATALOGING-IN-PUBLICATION DATA

Alternative energy / Neil Schlager and Jayne Weisblatt, editors.
   p. cm.
      Includes bibliographical references and index.
      ISBN 0-7876-9440-1 (set hardcover : alk. paper) –
      ISBN 0-7876-9439-8 (vol 1 : alk. paper) –
      ISBN 0-7876-9441-X (vol 2 : alk. paper) –
      ISBN 0-7876-9442-8 (vol 3 : alk. paper)
      1. Renewable energy sources. I. Schlager, Neil, 1966- II. Weisblatt, Jayne.

TJ808.A475 2006
333.79'4–dc22   2006003763

This title is also available as an e-book
ISBN 1-4414-0507-3
Contact your Thomson Gale sales representative for ordering information.

Printed in China
10 9 8 7 6 5 4 3 2 1

■■■

Contents

CHAPTER 8: WIND ENERGY

CHAPTER 9: ENERGY CONSERVATION AND EFFICIENCY

CHAPTER 10: POSSIBLE FUTURE ENERGY SOURCES

Introduction

Alternative Energy offers readers comprehensive and easy-to-use information on the development of alternative energy sources. Although the set focuses on new or emerging energy sources, such as geothermal power and solar energy, it also discusses existing energy sources such as those that rely on fossil fuels. Each volume begins with a general overview that presents the complex issues surrounding existing and potential energy sources. These include the increasing need for energy, the world's current dependence on nonrenewable sources of energy, the impact on the environment of current energy sources, and implications for the future. The overview will help readers place the new and alternative energy sources in perspective.

Each of the first eight chapters in the set covers a different energy source. These chapters each begin with an overview that defines the source, discusses its history and the scientists who developed it, and outlines the applications and technologies for using the source. Following the chapter overview, readers will find information about specific technologies in use and potential uses as well. Two additional chapters explore the need for conservation and the move toward more energy-efficient tools, building materials, and vehicles and the more theoretical (and even imaginary) energy sources that might become reality in the future.

ADDITIONAL FEATURES

Each volume of *Alternative Energy* includes the overview, a glossary called "Words to Know," a list of sources for more information, and an index. The set has 100 photos, charts, and illustrations to enliven the text, and sidebars provide additional facts and related information.

ACKNOWLEDGEMENTS

U•X•L would like to thank several individuals for their assistance with this set. At Schlager Group, Jayne Weisblatt and Neil Schlager oversaw the writing and editing of the set. Michael J. O'Neal, Amy Hackney Blackwell, and A. Petruso wrote the text for the volumes.

In addition, U•X•L editors would like to thank Dr. Peter Brimblecombe for his expert review of these volumes. Dr. Brimblecombe teaches courses on air pollution at the School of Environmental Sciences, University of East Anglia, United Kingdom. The editors also express their thanks for last minute contributions, review, and revisions to the final chapter on alternative and potential energy resources to Rory Clarke (physicist, CERN), Lee Wilmoth Lerner (electrical engineer and intern, NASA and the Fusion Research Laboratory at Auburn University), Larry Gilman (electrical engineer), and K. Lee Lerner (physicist and managing director, Lerner & Lerner, LLC).

COMMENTS AND SUGGESTIONS

We welcome your comments on *Alternative Energy* and suggestions for future editions of this work. Please write: Editors, *Alternative Energy*, U•X•L, 27500 Drake Rd., Farmington Hills, Michigan 48331-3535; call toll free: 1-800-877-4253; fax: 248-699-8097; or send e-mail via www.gale.com.

Words to Know

A

acid rain: Rain with a high concentration of sulfuric acid, which can damage cars, buildings, plants, and water supplies where it falls.

adobe: Bricks that are made from clay or earth, water, and straw, and dried in the sun.

alkane: A kind of hydrocarbon in which the molecules have the maximum possible number of hydrogen atoms and no double bonds.

anaerobic: Without air; in the absence of air or oxygen.

anemometer: A device used to measure wind speed.

anthracite: A hard, black coal that burns with little smoke.

aquaculture: The formal cultivation of fish or other aquatic life forms.

atomic number: The number of protons in the nucleus of an atom.

atomic weight: The combined number of an atom's protons and neutrons.

attenuator: A device that reduces the strength of an energy wave, such as sunlight.

B

balneology: The science of bathing in hot water.

barrel: A common unit of measurement of crude oil, equivalent to 42 U.S. gallons; barrels of oil per day, or BOPD, is a standard measurement of how much crude oil a well produces.

biodiesel: Diesel fuel made from vegetable oil.

bioenergy: Energy produced through the combustion of organic materials that are constantly being created, such as plants.

biofuel: A fuel made from organic materials that are constantly being created.

biomass: Organic materials that are constantly being created, such as plants.

bitumen: A black, viscous (oily) hydrocarbon substance left over from petroleum refining, often used to pave roads.

bituminous coal: Mid-grade coal that burns with a relatively high flame and smoke.

brine: Water that is very salty, such as the water found in the ocean.

British thermal unit (Btu or BTU): A measure of heat energy, equivalent to the amount of energy it takes to raise the temperature of one pound of water by one degree Fahrenheit.

butyl rubber: A synthetic rubber that does not easily tear. It is often used in hoses and inner tubes.

C

carbon sequestration: Storing the carbon emissions produced by coal-burning power plants so that pollutants are not released in the atmosphere.

catalyst: A substance that speeds up a chemical reaction or allows it to occur under different conditions than otherwise possible.

cauldron: A large metal pot.

CFC (chlorofluorocarbon): A chemical compound used as a refrigerant and propellant before being banned for fear it was destroying the ozone layer.

Clean Air Act: A U.S. law intended to reduce and control air pollution by setting emissions limits for utilities.

climate-responsive building: A building, or the process of constructing a building, using materials and techniques that take advantage of natural conditions to heat, cool, and light the building.

coal: A solid hydrocarbon found in the ground and formed from plant matter compressed for millions of years.

coke: A solid organic fuel made by burning off the volatile components of coal in the absence of air.

cold fusion: Nuclear fusion that occurs without high heat; also referred to as low energy nuclear reactions.

combustion: Burning.

compact fluorescent bulb: A lightbulb that saves energy as conventional fluorescent bulbs do, but that can be used in fixtures that normally take incandescent lightbulbs.

compressed: To make more dense so that a substance takes up less space.

conductive: A material that can transmit electrical energy.

convection: The circulation movement of a substance resulting from areas of different temperatures and/or densities.

core: The center of the Earth.

coriolis force: The movement of air currents to the right or left caused by Earth's rotation.

corrugated steel: Steel pieces that have parallel ridges and troughs.

critical mass: An amount of fissile material needed to produce an ongoing nuclear chain reaction.

criticality: The point at which a nuclear fission reaction is in controlled balance.

crude oil: The unrefined petroleum removed from an oil well.

crust: The outermost layer of the Earth.

curie: A unit of measurement that measures an amount of radiation.

current: The flow of electricity.

D

decay: The breakdown of a radioactive substance over time as its atoms spontaneously give off neutrons.

deciduous trees: Trees that shed their leaves in the fall and grow them in the spring. Such trees include maples and oaks.

decommission: To take a nuclear power plant out of operation.

dependent: To be reliant on something.

distillation: A process of separating or purifying a liquid by boiling the substance and then condensing the product.

distiller's grain: Grain left over from the process of distilling ethanol, which can be used as inexpensive high-protein animal feed.

drag: The slowing force of the wind as it strikes an object.

drag coefficient: A measurement of the drag produced when an object such as a car pushes its way through the air.

E

E85:   A blend of 15 percent ethanol and 85 percent gasoline.

efficient: To get a task done without much waste.

electrolysis: A method of producing chemical energy by passing an electric current through a type of liquid.

electromagnetism: Magnetism developed by a current of electricity.

electron: A negatively charged particle that revolves around the nucleus in an atom.

embargo: Preventing the trade of a certain type of commodity.

emission: The release of substances into the atmosphere. These substances can be gases or particles.

emulsion: A liquid that contains many small droplets of a substance that cannot dissolve in the liquid, such as oil and water shaken together.

enrichment: The process of increasing the purity of a radioactive element such as uranium to make it suitable as nuclear fuel.

ethanol: An alcohol made from plant materials such as corn or sugar cane that can be used as fuel.

experimentation: Scientific tests, sometimes of a new idea.

F

feasible: To be possible; able to be accomplished or brought about.

feedstock: A substance used as a raw material in the creation of another substance.

field: An area that contains many underground reservoirs of petroleum or natural gas.

fissile: Term used to describe any radioactive material that can be used as fuel because its atoms can be split.

fission: Splitting of an atom.

flexible fuel vehicle (FFV):   A vehicle that can run on a variety of fuel types without modification of the engine.

flow: The volume of water in a river or stream, usually expressed as gallons or cubic meters per unit of time, such as a minute or second.

fluorescent lightbulb: A lightbulb that produces light not with intense heat but by exciting the atoms in a phosphor coating inside the bulb.

fossil fuel: An organic fuel made through the compression and heating of plant matter over millions of years, such as coal, petroleum, and natural gas.

fusion: The process by which the nuclei of light atoms join, releasing energy.

G

gas: An air-like substance that expands to fill whatever container holds it, including natural gas and other gases commonly found with liquid petroleum.

gasification: A process of converting the energy from a solid, such as coal, into gas.

gasohol: A blend of gasoline and ethanol.

gasoline: Refined liquid petroleum most commonly used as fuel in internal combustion engines.

geothermal: Describing energy that is found in the hot spots under the Earth; describing energy that is made from heat.

geothermal reservoir: A pocket of hot water contained within the Earth's mantle.

global warming: A phenomenon in which the average temperature of the Earth rises, melting icecaps, raising sea levels, and causing other environmental problems.

gradient: A gradual change in something over a specific distance.

green building: Any building constructed with materials that require less energy to produce and that save energy during the building's operation.

greenhouse effect: A phenomenon in which gases in the Earth's atmosphere prevent the sun's radiation from being reflected back into space, raising the surface temperature of the Earth.

greenhouse gas: A gas, such as carbon dioxide or methane, that is added to the Earth's atmosphere by human actions. These gases trap heat and contribute to global warming.

H

halogen lamp: An incandescent lightbulb that produces more light because it produces more heat, but lasts longer because the filament is enclosed in quartz.

Heisenberg uncertainty principle: The principle that it is impossible to know simultaneously both the location and momentum of a subatomic particle.

heliostat: A mirror that reflects the sun in a constant direction.

hybrid vehicle: Any vehicle that is powered in a combination of two ways; usually refers to vehicles powered by an internal combustion engine and an electric motor.

hybridized: The bringing together of two different types of technology.

hydraulic energy: The kinetic energy contained in water.

hydrocarbon: A substance composed of the elements hydrogen and carbon, such as coal, petroleum, and natural gas.

hydroelectric: Describing electric energy made by the movement of water.

hydropower: Any form of power derived from water.

I

implement: To put something into practice.

incandescent lightbulb: A conventional lightbulb that produces light by heating a filament to high temperatures.

infrastructure: The framework that is necessary to the functioning of a structure; for example, roads and power lines form part of the infrastructure of a city.

inlet: An opening through which liquid enters a device, or place.

internal combustion engine: The type of engine in which the burning that generates power takes place inside the engine.

isotope: A "species" of an element whose nucleus contains more neutrons than other species of the same element.

K

kilowatt-hour: One kilowatt of electricity consumed over a one-hour period.

kinetic energy: The energy associated with movement, such as water that is in motion.

Kyoto Protocol: An international agreement among many nations setting limits on emissions of greenhouse gases; intended to slow or prevent global warming.

L

lava: Molten rock contained within the Earth that emerges from cracks in the Earth's crust, such as volcanoes.

lift: The aerodynamic force that operates perpendicular to the wind, owing to differences in air pressure on either side of a turbine blade.

lignite: A soft brown coal with visible traces of plant matter in it that burns with a great deal of smoke and produces less heat than anthracite or bituminous coal.

liquefaction: The process of turning a gas or solid into a liquid.

LNG (liquefied natural gas): Gas that has been turned into liquid through the application of pressure and cold.

LPG (liquefied petroleum gas): A gas, mainly propane or butane, that has been turned into liquid through the use of pressure and cold.

lumen: A measure of the amount of light, defined as the amount of light produced by one candle.

M

magma: Liquid rock within the mantle.

magnetic levitation: The process of using the attractive and repulsive forces of magnetism to move objects such as trains.

mantle: The layer of the Earth between the core and the crust.

mechanical energy: The energy output of tools or machinery.

meltdown: Term used to refer to the possibility that a nuclear reactor could become so overheated that it would melt into the earth below.

mica: A type of shiny silica mineral usually found in certain types of rocks.

modular: An object which can be easily arranged, rearranged, replaced, or interchanged with similar objects.

mousse: A frothy mixture of oil and seawater in the area where an oil spill has occurred.

N

nacelle: The part of a wind turbine that houses the gearbox, generator, and other components.

natural gas: A gaseous hydrocarbon commonly found with petroleum.

negligible: To be so small as to be insignificant.

neutron: A particle with no electrical charge found in the nucleus of most atoms.

NGL (natural gas liquid): The liquid form of gases commonly found with natural gas, such as propane, butane, and ethane.

nonrenewable: To be limited in quantity and unable to be replaced.

nucleus: The center of an atom, containing protons and in the case of most elements, neutrons.

O

ocean thermal energy conversion (OTEC): The process of converting the heat contained in the oceans' water into electrical energy.

octane rating: The measure of how much a fuel can be compressed before it spontaneously ignites.

off-peak: Describing period of time when energy is being delivered at well below the maximum amount of demand, often nighttime.

oil: Liquid petroleum; a substance refined from petroleum used as a lubricant.

organic: Related to or derived from living matter, such as plants or animals; composed mainly of carbon atoms.

overburden: The dirt and rocks covering a deposit of coal or other fossil fuel.

oxygenate: A substance that increases the oxygen level in another substance.

ozone: A molecule consisting of three atoms of oxygen, naturally produced in the Earth's atmosphere; ozone is toxic to humans.

P

parabolic: Shaped like a parabola, which is a certain type of curve.

paraffin: A kind of alkane hydrocarbon that exists as a white, waxy solid at room temperature and can be used as fuel or as a wax for purposes such as sealing jars or making candles.

passive: A device that takes advantage of the sun's heat but does not use an additional source of energy.

peat: A brown substance composed of compressed plant matter and found in boggy areas; peat can be used as fuel itself, or turns into coal if compressed for long enough.

perpetual motion: The power of a machine to run indefinitely without any energy input.

petrochemicals: Chemical compounds that form in rocks, such as petroleum and coal.

petrodiesel: Diesel fuel made from petroleum.

petroleum: Liquid hydrocarbon found underground that can be refined into gasoline, diesel fuel, oils, kerosene, and other products.

pile: A mass of radioactive material in a nuclear reactor.

plutonium: A highly toxic element that can be used as fuel in nuclear reactors.

polymer: A compound, either synthetic or natural, that is made of many large molecules. These molecules are made from smaller, identical molecules that are chemically bonded.

pristine: Not changed by human hands; in its original condition.

productivity: The output of labor per amount of work.

proponent: Someone who supports an idea or cause.

proton: A positively charged particle found in the nucleus of an atom.

R

radioactive: Term used to describe any substance that decays over time by giving off subatomic particles such as neutrons.

RFG (reformulated gasoline): Gasoline that has an oxygenate or other additive added to it to decrease emissions and improve performance.

rem: An abbreviation for "roentgen equivalent man," referring to a dose of radiation that will cause the same biological effect (on a "man") as one roentgen of X-rays or gamma rays.

reservoir: A geologic formation that can contain liquid petroleum and natural gas.

reservoir rock: Porous rock, such as limestone or sandstone, that can hold accumulations of petroleum or natural gas.

retrofit: To change something, like a home, after it is built.

rotor: The hub to which the blades of a wind turbine are connected; sometimes used to refer to the rotor itself and the blades as a single unit.

S

scupper: An opening that allows a liquid to drain.

seam: A deposit of coal in the ground.

sedimentary rock: A rock formed through years of minerals accumulating and being compressed.

seismology: The study of movement within the earth, such as earthquakes and the eruption of volcanoes.

sick building syndrome: The tendency of buildings that are poorly ventilated, lighted, and humidified, and that are made with certain synthetic materials to cause the occupants to feel ill.

smog: Air pollution composed of particles mixed with smoke, fog, or haze in the air.

stall: The loss of lift that occurs when a wing presents too steep an angle to the wind and low pressure along the upper surface of the wing decreases.

strip mining: A form of mining that involves removing earth and rocks by bulldozer to retrieve the minerals beneath them.

stored energy: The energy contained in water that is stored in a tank or held back behind a dam in a reservoir.

subsidence: The collapse of earth above an empty mine, resulting in a damaged landscape.

surcharge: An additional charge over and above the original cost.

superconductivity: The disappearance of electrical resistance in a substance such as some metals at very low temperatures.

T

thermal energy: Any form of energy in the form of heat; used in reference to heat in the oceans' waters.

thermal gradient: The differences in temperature between different layers of the oceans.

thermal mass: The measure of the amount of heat a substance can hold.

thermodynamics: The branch of physics that deals with the mechanical actions or relations of heat.

tokamak: An acronym for the Russian-built toroidal magnetic chamber, a device for containing a fusion reaction.

transitioning: Changing from one position or state to another.

transparent: So clear that light can pass through without distortion.

trap: A reservoir or area within Earth's crust made of nonporous rock that can contain liquids or gases, such as water, petroleum, and natural gas.

trawler: A large commercial fishing boat.

Trombé wall: An exterior wall that conserves energy by trapping heat between glazing and a thermal mass, then venting it into the living area.

turbine: A device that spins to produce electricity.

U

uranium: A heavy element that is the chief source of fuel for nuclear reactors.

V

viable: To be possible; to be able to grow or develop.

voltage: Electric potential that is measured in volts.

W

wind farm: A group of wind turbines that provide electricity for commercial uses.

work: The conversion of one form of energy into another, such as the conversion of the kinetic energy of water into mechanical energy used to perform a task.

Z

zero point energy: The energy contained in electromagnetic fluctuations that remains in a vacuum, even when the temperature has been reduced to very low levels.

▪▪▪
Overview

In the technological world of the twenty-first century, few people can truly imagine the challenges faced by prehistoric people as they tried to cope with their natural environment. Thousands of years ago life was a daily struggle to find, store, and cook food, stay warm and clothed, and generally survive to an "old age" equal to that of most of today's college students. A common image of prehistoric life is that of dirty and ill-clad people huddled around a smoky campfire outside a cave in an ongoing effort to stay warm and dry and to stop the rumbling in their bellies.

The "caves" of the twenty-first century are a little cozier. The typical person, at least in more developed countries, wakes up each morning in a reasonably comfortable house because the gas, propane, or electric heating system (or electric air-conditioner) has operated automatically overnight. A warm shower awaits because of hot water heaters powered by electricity or natural gas, and hair dries quickly (and stylishly) under an electric hair dryer. An electric iron takes the wrinkles out of the clean shirt that sat overnight in the electric clothes dryer. Milk for a morning bowl of cereal remains fresh in an electric refrigerator, and it costs pennies per bowl thanks to electrically powered milking operations on modern dairy farms. The person then goes to the garage (after turning off all the electric lights in the house), hits the electric garage door opener, and gets into his or her gasoline-powered car for the drive to work—perhaps in an office building that consumes power for lighting, heating and air-conditioning, copiers, coffeemakers, and computers. Later, an electric, propane, or natural gas stove is used to cook dinner. Later still, an electric popcorn popper provides a snack as the person watches an electric television or reads under the warm glow of electric light bulbs— after perhaps turning up the heat because the house is a little chilly.

CATASTROPHE AHEAD?

Most people take these modern conveniences for granted. Few people give much thought to them, at least until there is a power outage or prices rise sharply, as they did for gasoline in the United States in the summer and fall of 2005. Many scientists, environmentalists, and concerned members of the public, though, believe that these conveniences have been taken too much for granted. Some believe that the modern reliance on fossil fuels—fuels such as natural gas, gasoline, propane, and coal that are processed from materials mined from the earth—has set the Earth on a collision course with disaster in the twenty-first century. Their belief is that the human community is simply burning too much fuel and that the consequences of doing so will be dire (terrible). Some of their concerns include the following:

- Too much money is spent on fossil fuels. In the United States, over $1 billion is spent every day to power the country's cars and trucks.
- Much of the supply of fossil fuels, particularly petroleum, comes from areas of the world that may be unstable. The U.S. fuel supply could be cut off without warning by a foreign government. Many nations that import all or most of their petroleum feel as if they are hostages to the nations that control the world's petroleum supplies.
- Drilling for oil and mining coal can do damage to the landscape that is impossible to repair.
- Reserves of coals and especially oil are limited, and eventually supplies will run out. In the meantime, the cost of such fuels will rise dramatically as it becomes more and more difficult to find and extract them.
- Transporting petroleum in massive tankers at sea heightens the risk of oil spills, causing damage to the marine and coastal environments.

Furthermore, to provide heat and electricity, fossil fuels have to be burned, and this burning gives rise to a host of problems. It releases pollutants in the form of carbon dioxide and sulfur into the air, fouling the atmosphere and causing "brown clouds" over cities. These pollutants can increase health problems such as lung disease. They may also contribute to a phenomenon called "global warming." This term refers to the theory that average temperatures across the globe will increase as "greenhouse gases" such as carbon dioxide trap the sun's heat (as a greenhouse does) in the atmosphere and warm it. Global warming, in turn, can melt glaciers and the polar ice caps, raising sea levels with damaging effects on coastal cities and small island nations. It may also cause climate changes, crop failures, and more unpredictable weather patterns.

Some scientists do not believe that global warming even exists or that its consequences will be catastrophic. Some note that throughout history, the world's average temperatures have risen and fallen. Some do not find the scientific data about temperature, glacial melting, rising sea levels, and unpredictable weather totally believable. While the debate continues, scientists struggle to learn more about the effects of human activity on the environment. At the same time, governments struggle to maintain a balance between economic development and its possible effects on the environment.

WHAT TO DO?

These problems began to become more serious after the Industrial Revolution of the nineteenth century. Until that time people depended on other sources of power. Of course, they burned coal or wood in fireplaces and stoves, but they also relied on the power of the sun, the wind, and river currents to accomplish much of their work. The Industrial Revolution changed that. Now, coal was being burned in vast amounts to power factories and steam engines as the economies of Europe and North America grew and developed. Later, more efficient electricity became the preferred power source, but coal still had to be burned to produce electricity in large power plants. Then in 1886 the first internal combustion engine was developed and used in an automobile. Within a few decades there was a demand for gasoline to power these engines. By 1929 the number of cars in the United States had grown to twenty-three million, and in the quarter-century between 1904 and 1929, the number of trucks grew from just seven hundred to 3.4 million.

At the same time technological advances improved life in the home. In 1920, for example, the United States produced a total of five thousand refrigerators. Just ten years later the number had grown to one million per year. These and many other industrial and consumer developments required vast and growing amounts of fuel. Compounding the problem in the twenty-first century is that other nations of the world, such as China and India, have started to develop more modern industrialized economies powered by fossil fuels.

By the end of World War II in 1945, scientists were beginning to imagine a world powered by fuel that was cheap, clean, and inexhaustible (unable to be used up). During the war the United States had unleashed the power of the atom to create the atomic bomb. Scientists believed that the atom could be used for peaceful purposes in nuclear power plants. They even envisioned (imagined) a day when homes could be powered by their own tiny nuclear power generators. This dream proved to be just that. While some four hundred nuclear power plants worldwide provide about 16 percent of the world's electricity, building such plants is an enormously expensive technical feat. Moreover, nuclear power plants produce spent fuel that is dangerous and not easily disposed of. The public fears that an accident at such a plant could release deadly radiation that would have disastrous effects on the surrounding area. Nuclear power has strong defenders, but it is not cheap, and safety concerns sometimes make it unpopular.

The dream of a fuel source that is safe, plentiful, clean, and inexpensive, however, lives on. The awareness of the need for such alternative fuel sources became greater in the 1970s, when the oil-exporting countries of the Middle East stopped shipments of oil to the United States and its allies. This situation (an embargo) caused fuel shortages and rapidly rising prices at the gas pump. In the decades that followed, gasoline again became plentiful and relatively inexpensive, but the oil embargo served as a wakeup call for many people. In addition, during these years people worldwide grew concerned about pollution, industrialization, and damage to the environment. Accordingly, efforts were intensified to find and develop alternative sources of energy.

ALTERNATIVE ENERGY: BACK TO THE FUTURE

Some of these alternative fuel sources are by no means new. For centuries people have harnessed the power of running water for a variety of needs, particularly for agriculture (farming). Water wheels were constructed in the Middle East, Greece, and China thousands of years ago, and they were common fixtures on the farms of Europe by the Middle Ages. In the early twenty-first century hydroelectric dams, which generate electricity from the power of rivers, provide about 9 percent of the electricity in the

United States. Worldwide, there are about 40,000 such dams. In some countries, such as Norway, hydroelectric dams provide virtually 100 percent of the nation's electrical needs. Scientists, though, express concerns about the impact such dams have on the natural environment.

Water can provide power in other ways. Scientists have been attempting to harness the enormous power contained in ocean waves, tides, and currents. Furthermore, they note that the oceans absorb enormous amounts of energy from the sun, and they hope someday to be able to tap into that energy for human needs. Technical problems continue to occur. It remains likely that ocean power will serve only to supplement (add to) existing power sources in the near future.

Another source of energy that is not new is solar power. For centuries, people have used the heat of the sun to warm houses, dry laundry, and preserve food. In the twenty-first century such "passive" uses of the sun's rays have been supplemented with photovoltaic devices that convert the energy of the sun into electricity. Solar power, though, is limited geographically to regions of the Earth where sunshine is plentiful.

Another old source of heat is geothermal power, referring to the heat that seeps out of the earth in places such as hot springs. In the past this heat was used directly, but in the modern world it is also used indirectly to produce electricity. In 1999 over 8,000 megawatts (that is, 8,000 million watts) of electricity were produced by about 250 geothermal power plants in twenty-two countries around the world. That same year the United States produced nearly 3,000 megawatts of geothermal electricity, more than twice the amount of power generated by wind and solar power. Geothermal power, though, is restricted by the limited number of suitable sites for tapping it.

Finally, wind power is getting a closer look. For centuries people have harnessed the power of the wind to turn windmills, using the energy to accomplish work. In the United States, wind-operated turbines produce just 0.4 percent of the nation's energy needs. However, wind experts believe that a realistic goal is for wind to supply 20 percent of the nation's electricity requirements by 2020. Worldwide, wind supplies enough power for about nine million homes. Its future development, though, is hampered by limitations on the number of sites with enough wind and by concerns about large numbers of unsightly wind turbines marring the landscape.

ALTERNATIVE ENERGY: FORWARD TO THE FUTURE

While some forms of modern alternative energy sources are really developments of long-existing technologies, others are genuinely new, though scientists have been exploring even some of these for up to hundreds of years. One, called bioenergy, refers to the burning of biological materials that otherwise might have just been thrown away or never grown in the first place. These include animal waste, garbage, straw, wood by-products, charcoal, dried plants, nutshells, and the material left over after the processing of certain foods, such as sugar and orange juice. Bioenergy also includes methane gas given off by garbage as it decomposes or rots. Fuels made from vegetable oils can be used to power engines, such as those in cars and trucks. Biofuels are generally cleaner than fossil fuels, so they do not pollute as much, and they are renewable. They remain expensive, and amassing significant amounts of biofuels requires a large commitment of agricultural resources such as farmland.

Nothing is sophisticated about burning garbage. A more sophisticated modern alternative is hydrogen, the most abundant element in the universe. Hydrogen in its pure form is extremely flammable. The problem with using hydrogen as a fuel is separating hydrogen molecules from the other elements to which it readily bonds, such as oxygen (hydrogen and oxygen combine to form water). Hydrogen can be used in fuel cells, where water is broken down into its elements. The hydrogen becomes fuel, while the "waste product" is oxygen. Many scientists regard hydrogen fuel cells as the "fuel of the future," believing that it will provide clean, safe, renewable fuel to power homes, office buildings, and even cars and trucks. However, fuel cells are expensive. As of 2002 a fuel cell could cost anywhere from $500 to $2,500 per kilowatt produced. Engines that burn gasoline cost only about $30 to $35 for the same amount of energy.

All of these power sources have high costs, both for the fuel and for the technology needed to use it. The real dreamers among energy researchers are those who envision a future powered by a fuel that is not only clean, safe, and renewable but essentially free. Many scientists believe that such fuel alternatives are impossible, at least for the foreseeable future. Others, though, work in laboratories around the world to harness more theoretical sources of energy. Some of their work has a "science fiction" quality, but these scientists point out that a few hundred years ago the airplane was science fiction.

One of these energy sources is magnetism, already used to power magnetic levitation ("maglev") trains in Japan and Germany. Another is perpetual motion, the movement of a machine that produces energy without requiring energy to be put into the system. Most scientists, though, dismiss perpetual motion as a violation of the laws of physics. Other scientists are investigating so-called zero-point energy, or the energy that surrounds all matter and can even be found in the vacuum of space. But perhaps the most sought-after source of energy for investigators is cold fusion, a nuclear reaction using "heavy hydrogen," an abundant element in seawater, as fuel. With cold fusion, power could be produced literally from a bucket of water. So far, no one has been able to produce it, though some scientists claim to have come very close.

None of these energy sources is a complete cure for the world's energy woes. Most will continue to serve as supplements to conventional fossil fuel burning for decades to come. But with the commitment of research dollars, it is possible that future generations will be able to generate all their power needs in ways that scientists have not even yet imagined. The first step begins with understanding fossil fuels, the energy they provide, the problems they cause, and what it may take to replace them.

Fossil Fuels

INTRODUCTION: WHAT ARE FOSSIL FUELS?

Nearly 90 percent of the world's energy comes from fossil fuels. Because fossil fuels are the main source, they are not alternative energy sources. Fossil fuels include coal, natural gas, and petroleum (puh-TROH-lee-uhm), which is often called oil. People use fossil fuels to meet nearly all of their energy needs, such as powering cars, producing electricity for light and heat, and running factories. Because their use is so widespread, it is important to understand fossil fuels in order to make informed decisions about present and future alternative energy sources.

Fossil fuels are a popular source of energy because they are considered convenient, effective, plentiful, and inexpensive, but a few nations have most of the world's fossil fuels, a fact that often causes conflicts. Nevertheless, as of 2006, there are no practical and available alternatives to fossil fuels for most energy needs, so they continue to be heavily used.

Types of fossil fuels

Fossil fuels are substances that formed underground millions of years ago from prehistoric plants and other living things that were buried under layers of sediment, which included dirt, sand, and dead plants. To turn into fossil fuels, this organic matter (matter that comes from a life form and is composed mainly of the element carbon) was crushed, heated, and deprived of oxygen. Under the right conditions and over millions of years, this treatment turns dead plants into fossil fuels.

The three main types of fossil fuels correspond to the three states of matter—solid, liquid, and gas:

- Coal is a solid.

Words to Know

Alkane A kind of hydrocarbon in which the molecules have the maximum possible number of hydrogen atoms and no double bonds.

Barrel A common unit of measurement of crude oil, equivalent to 42 U.S. gallons; barrels of oil per day, or BOPD, is a standard measurement of how much crude oil a well produces.

Catalyst A substance that speeds up a chemical reaction or allows it to occur under different conditions than otherwise possible.

Clean Air Act A U.S. law intended to reduce and control air pollution by setting emissions limits for utilities.

Emissions The by-products of fossil fuel burning that are released into the air.

Global warming A phenomenon in which the average temperature of the Earth rises, melting icecaps, raising sea levels, and causing other environmental problems.

Greenhouse effect A phenomenon in which gases in the Earth's atmosphere prevent the sun's radiation from being reflected back into space, raising the surface temperature of the Earth.

Octane rating The measure of how much a fuel can be compressed before it spontaneously ignites.

Ozone A molecule consisting of three atoms of oxygen, naturally produced in the Earth's atmosphere; ozone is toxic to humans.

Seismology The study of movement within the earth, such as earthquakes and the eruption of volcanoes.

- Petroleum is a liquid.
- Natural gas is a gas.

Several fossil fuels are made by refining petroleum or natural gas. These fuels include gases such as propane, butane, and methanol.

Natural Gas Versus Gasoline

Natural gas is not sold at gas stations. The fuel used in cars is liquid petroleum, or gasoline. Although most people call it "gas," this fuel is not the same thing as natural gas. The word *gas* refers to natural gas, not gasoline. The word *oil* refers to petroleum.

Whether a fossil fuel formed as a solid, liquid, or gas depends on the location, the composition of the materials, the length of time the matter was compressed, how hot it became, and how long it was buried. Coal formed from accumulated layers of plants that died in swamps and were buried for millions of years. Petroleum and natural gas formed from microscopic plants and bacteria in the oceans. Both petroleum and natural gas formed in places that could contain them: pockets, or reservoirs (reh-zuh-VWARS), in the undersea rock.

Dinosaurs in the Gas Tank

It is unlikely that fossil fuels are made of dinosaurs. Most fossil fuels formed about 300 million years ago, and most of them are made mainly of plant matter. Dinosaurs did not appear until about 230 million years ago, so the first dinosaur was not born until the youngest petroleum had already formed. Dinosaur fossils, however, do have something in common with fossil fuels. Fossils, whether they are dinosaurs or coal, are the hardened remains of animals and plants preserved in Earth's crust from an earlier age. Dinosaur fossils formed when dinosaurs were buried in sand or dirt, and their skeletons were hardened by minerals that seeped in through tiny holes in the bone.

Earth has a lot of fossil fuels. Scientists in 2005 estimated that the ground contains about ten trillion metric tons of coal, enough to fuel human energy needs for hundreds of years. Petroleum and natural gas deposits are not nearly so extensive. Most scientists believe that if people keep using up oil and gas at 2005 rates, all known petroleum and gas reserves will be used up by the beginning of the twenty-second century.

At the end of the twentieth century, petroleum supplied about 40 percent of the energy needs of the United States. Another 22 percent was covered by coal and 24 percent by natural gas. The International Energy Agency (IEA) has predicted that the world will need almost 60 percent more energy in 2030 than it did in 2002. The IEA believes that fossil fuels will still be supplying most of those needs by 2030.

Other kinds of fossil fuels exist, but none of them can be extracted, recovered, or used efficiently. These fossil fuels include:

- Gas hydrates, which are deposits of methane and water that form crystals in ocean sediments. There is currently no technology for extracting methane from the crystals, so gas hydrates are not yet considered a part of world energy reserves.
- Tar sands, which are patches of tar in sandstone. Petroleum sometimes gets embedded in sandstone, and the bacteria in the sandstone and the surrounding water make the petroleum turn into tar. Tar sands are difficult to recover and use.

- Oil shale, which is a kind of rock full of a waxy organic substance called kerogen (KEHR-uh-juhn). Kerogen formed from the same microscopic plants and bacteria that make up petroleum, but it never reached the pressure or temperature that would have turned it into oil. It is not currently practical to recover or use oil shale.

How fossil fuels work

Fossil fuels generate energy by burning. This energy can serve a variety of purposes from heating homes to powering automobiles. The simplest devices that use fossil fuels burn them so that people can take advantage of the heat. For example, some homes are heated by furnaces that burn natural gas. The heat from the burning gas warms the house. Camping stoves often burn propane that is fed to the stove burners from an attached bottle. Coal stoves burn lumps of coal.

Most fossil fuel-powered operations, however, use the burning of the fossil fuel to power much more complex machines, such as internal combustion engines. In many cases, other fuels could supply the necessary heat; for example, locomotives could be powered by burning wood instead of burning coal, and power plants can be powered by water instead of coal. The advantage of fossil fuels in these situations is that they produce large amounts of heat for their volume, and they are currently widely available, with some liquid and gas fuels available at pumps.

The internal combustion engine

Automobiles use fossil fuel (gasoline) to power their internal combustion engines. An internal combustion engine burns a fuel to power pistons, which make the engine turn. Internal combustion engines have been around since the 1860s. The four-stroke "Otto" engine was invented in 1867 by Nikolaus August Otto (1832–1891), a German engineer. Another German engineer, Rudolph Diesel (1858–1913), invented the diesel engine in 1892. The basic principles of internal combustion have not changed since then.

An engine contains several cylinders (most cars have between four and eight) that make the engine move. A four-stroke cylinder works like this:

1. The intake valve opens to let air and fuel into the cylinder while the piston is down. This is called the intake stroke.

2. The piston begins traveling back up. The intake valve closes and the piston compresses the air and fuel in the cylinder. This is called the compression stroke.

3. The spark plug creates a spark, which ignites the fuel and air so that it explodes. The explosion pushes the piston down. The piston rotates the crankshaft, which turns the engine. This is called the power stroke.

4. The exhaust valve opens. The piston moves back up, forcing the burned gases out through the exhaust valve. The piston travels back down, the exhaust valve opens, and the intake stroke begins again. This is called the exhaust stroke.

One complete cycle of a four-stroke engine will turn the crankshaft twice. A car engine's cylinders can fire hundreds of times in a minute, turning the crankshaft, which transmits its energy into turning the car's wheels. The more air and fuel that can get into a cylinder, the more powerful the engine will be.

Photo of the original 1891 gasoline-engined Daimler automobile. In 1885, Karl Benz and Gottlieb Daimler developed an internal combustion engine, building the first motorcycle and cars using gasoline. © AP Images.

What does Octane Mean?

Gasoline comes in several varieties labeled with words such as "regular" or "supreme," each with a number. The higher the number on the gasoline, the more expensive it is. That number is the gasoline's octane rating, which tells how much the fuel can be compressed before it will spontaneously ignite. In a car engine, gasoline is supposed to ignite in one of the engine's cylinders when it is lit by a spark plug; it is not supposed to ignite on its own. When it ignites on its own, the engine "knocks." This can damage the engine. High-performance cars, though, increase their horsepower by increasing the amount of compression in the engine, which makes knocking more likely. That is why high-performance cars have to use expensive, high octane gasoline.

Most engines that run on gasoline can also be powered with natural gas or LPGs (liquefied petroleum gas), with some minor modifications to the fuel delivery system. The basic method of combustion is the same.

A diesel engine is similar to a gasoline engine except that only air enters the cylinder during the intake stroke, and only air is compressed during the compression stroke. The fuel is sprayed into the cylinder at the end of the compression stroke, when the air temperature is high enough to cause it to ignite spontaneously without a spark. Diesel engines are usually heavier and more powerful than gasoline engines and have better fuel efficiency; they are used in buses, trucks, ships, and some automobiles. In Europe, a large proportion of personal automobiles are powered by diesel fuel, but diesel fuel is less common in the United States because of clean-air laws. Diesel fuel has more exhaust emissions than gasoline.

Coal-burning engines

Using coal for heat and cooking can be as straightforward as putting coal in a stove and setting it on fire; the coal burns slowly and emits steady heat. But the way coal really had an effect on people's lives was through its use as a fuel for engines, such as steam engines that powered locomotives that pulled trains. Coal-burning locomotives used steam to power their wheels. A locomotive works like this:

Side view of George Stephenson's *Rocket* locomotive. The train was designed and built in 1829 and is considered the forerunner of all other steam locomotives. © *Hulton-Deutsch Collection/Corbis.*

1. In order to keep the fire burning, the locomotive has to carry a large pile of coal, which a person called the fireman constantly shovels into the firebed. (More modern locomotives have mechanical shovels to feed the fires.)

2. The ashes left over from the burning coal fall through grates into an ashpan below the firebed. The ashes are dumped at the end of the train's run.

3. This basic process was not only used in trains. Steam engines also powered riverboats, steamships, and factories.

Most trains in the twenty-first century are powered by diesel fuel or electricity. China still uses coal-burning trains for normal transportation, but in Europe and the United States steam locomotives are only used as part of museum displays to entertain tourists.

Where electricity comes from

Fossil fuels are important for the production of electricity. Most power plants have generators that spin to create electricity, which is then sent out through the wires and poles that distribute it to consumers. Something has to power those generators. The vast majority of power plants burn fossil fuels for this purpose. (The rest use nuclear power or hydroelectric power.)

About one-half of the electricity in the United States comes from coal-burning power plants. These plants store their coal in giant outdoor piles. People driving bulldozers push the coal onto conveyor belts that carry it up to silos or bunkers. The coal is typically crushed so it can be fed into most power station furnaces. Then it is fed into giant burners that burn night and day to create steam to turn the generator. Most plants need constant deliveries of coal to have enough fuel to keep the burners running at all times. They produce large amounts of ash. One of the jobs of plant operators is to keep the ash from clogging up the works.

Natural gas is the other significant fossil fuel source of electrical power in the United States, supplying about one-fifth of the nation's electricity. Natural gas plants use turbines to spin generators. The turbines are connected to pipelines that provide a constant supply of natural gas. Some plants use the natural gas to power the generator directly. Others use the natural gas to create steam, which spins the generator.

The United States government encourages power companies to build plants powered by natural gas because natural gas burns much more cleanly than coal and therefore does not create as much pollution. The U.S. Department of Energy predicts that 90 percent of new power plants built in the early 2000s will be powered by natural gas.

Historical overview: Notable discoveries and the people who made them

Humans have been using fossil fuels for thousands of years, possibly as long ago as twenty thousand years. Oil sometimes seeps up through the ground, so it was easy for people to see it and experiment with it. The ancient Mesopotamians in what is now Iraq may have discovered a way to use oil about five thousand years ago. Historians believe that people first used petroleum as oil for lighting, dipping wood in it and setting it on fire as a torch. Ancient Greeks and Romans used coal as a fuel for heat and cooking. Ancient temples sometimes had eternal flames, which may have been powered by natural gas leaking up from the ground.

In the British Isles coal began to be used in the late thirteenth century, and it was the dominant fuel in London by 1600. Wood was abundant, so coal took time to become widely adopted. The first widespread use of fossil fuels occurred in the late 1700s, with the

Illustration of the *Savannah*, the first steamship to cross the Atlantic Ocean.
© *Bettman/Corbis*.

development of the steam engine and the start of the industrial revolution. James Watt (1736–1819) is usually credited as the inventor of the first commercially efficient steam engine in 1769, though his work was based on the inventions of others, particularly that of the Cornish engineer Thomas Newcomen (1664–1729), whose atmospheric steam engine was completed in 1711. The steam engine, powered by coal, made the industrial revolution possible. Steam engines could power trains, boats, and factories. The first coal-burning steam locomotive was built in Wales in 1804. In 1825 coal-powered trains became available for commercial use. Robert Fulton (1765–1815) invented the steam-powered riverboat in 1807, and riverboats became a popular way to travel up and down the Mississippi River in the United States. In 1819 a steamship crossed the Atlantic Ocean for the first time. By the mid-1800s people were regularly traveling between Europe and the United States on coal-powered steamships.

People began using natural gas to power lamps in 1785 in England. Natural gas lamps became common in the United States around 1816. The first natural gas well was built in Fredonia, New York, in 1821.

In the 1850s an American lawyer named George Bissell (1821–1884) investigated the possibility of using oil as lamp fuel. He thought he could find more oil if he drilled into the ground, so he hired Edwin Drake (1819–1880) to drill the first oil well. This well was completed in Titusville, Pennsylvania, in 1859. Drake used the oil to make kerosene, which people used in lamps and heaters. Gasoline was a by-product (one of the leftovers) of the process of making kerosene, but no one at the time had a real use for it. Other people began looking for oil, and they found it in places such as Indonesia, Texas, and the Middle East.

By the end of the nineteenth century many people were using light bulbs instead of kerosene lamps, so oil producers began adapting their product for other uses. The first gasoline-powered internal combustion engine was developed in 1886. The first mass-produced gasoline-powered car was the Oldsmobile, introduced in 1902. Henry Ford (1863–1947) introduced the Model T in 1908 and began producing his inexpensive cars on an assembly line. By 1920 there were twenty-three million cars in the world, and it turned out that gasoline was the most practical way to power them.

The Wright brothers, Orville (1871–1948) and Wilbur (1867–1912), flew their first successful airplane in 1903. They used petroleum as their fuel, and from that point on airplanes were powered by petroleum-based fuels. Diesel fuel gradually replaced coal as the dominant fuel for large ships. Diesel locomotives appeared around 1920 and had replaced steam engines by 1960.

Consumption of all fossil fuels increased greatly during the twentieth century. Petroleum was used to power automobiles, airplanes, ships, and electric plants. Coal heated homes, powered factories and trains, and generated electricity at power plants. Toward the end of the twentieth century the oil industry began to develop the potential of natural gas, and this fuel became useful in homes and businesses as well as in industry. Minor fossil fuels such as kerosene, propane, and butane were all widely used at the beginning of the twenty-first century. Perhaps the most notable transition from the twentieth to the twenty-first century is from stationary devices burning solid fuels to mobile sources using liquid fuels.

Current and future technology

Fossil fuels supply a large percentage of the world's energy needs through a variety of technologies. Most automobiles and other vehicles use gasoline to power internal combustion engines, in which the burning that generates power takes place inside the engine. Coal or gasoline is burned to power factory equipment. Coal-fired plants generate much of the world's electricity. Almost every twenty-first century technology uses fossil fuels in some way.

Fossil fuel technology has changed. Scientists are constantly looking for technology that makes fossil fuels work more efficiently and reduces pollution. Fossil fuels are so common and considered so necessary that there is great incentive for engineers to improve methods of acquiring and using fossil fuels. Technology under development includes:

- Clean coal technology
- Vehicles powered by natural gas or substances other than gasoline
- Fuel cells that use small amounts of fossil fuel to make hydrogen
- Safer means of transporting fossil fuels
- Improved techniques for cleaning fossil fuels before, during, and after burning
- Improvements in extracting fossil fuels from the ground

Benefits and drawbacks of fossil fuels

Most existing technology was designed for use with fossil fuels. Fossil fuel transport systems are already in place. Pipelines for oil and natural gas and trucks and ships for petroleum products move the fossil fuels where they are needed. And consumers can buy the fossil fuel products they use on practically every corner.

Yet, fossil fuels are non-renewable resources. Current supplies took a very long time to form under the Earth's crust. These supplies will be gone long before the Earth has a chance to replace them. Even now, getting fossil fuels is a major drawback to using them. Countries that do not have reserves of oil and natural gas must depend on those countries that do. And using fossil fuels contributes to air and water pollution.

Environmental impact of fossil fuels

Fossil fuels cause or contribute to environmental problems such as the following:

- Damage to the landscape
- Air pollution
- Water pollution

- Oil spills
- Radioactivity (Coal contains the radioactive elements uranium and thorium, and most coal-fired plants emit more radiation than a nuclear power plant.)
- Health problems for workers and those nearby (Many fossil fuel byproducts can be harmful to humans: breathing toxic hydrocarbons, nitrogen oxides, and particulate matter can cause ailments such as chest pain, coughing, asthma, chronic bronchitis, decreased lung function, and cancer, and exposure to mercury can lead to nerve damage, birth defects, learning disabilities, and even death.)

Some experts believe the environmental problems are so serious that people need to find alternatives to fossil fuels even before all reserves are used up. Others believe that technological improvements will allow the use of fossil fuels for many years to come.

Damage to the landscape

Fossil fuels are found underground. There is no way to get them out without cutting into or removing the dirt on top of deposits. Strip mining for coal involves removing the dirt and rocks above a deposit of coal and digging out the coal beneath it. Miners sometimes remove the tops of mountains to remove the coal below. Mines below the Earth's surface can collapse, resulting in changes to the landscape on top of them.

Though drilling for oil and natural gas is not always as destructive as coal mining, it still involves machinery that can destroy animal habitats and pipelines that cut across the land for thousands of miles.

Air pollution

Air pollution results from driving cars and trucks, from burning coal and other fossil fuels to create electricity, from industry, from using gas-powered stoves and appliances, and from many other daily activities. As the number of drivers increases and average fuel efficiency declines due to a shift to lower mileage SUVs, air pollution increases. As the number of people using electricity increases, so does air pollution.

There are several types of air pollution:
- Particulate matter is tiny particles of burnt fossil fuels that float in the air. This kind of pollution is sometimes called black carbon pollution. Examples of coarse particulate matter include the smoke that comes from a diesel-powered

truck or the soot that rises from a charcoal-burning grill.
However, in addition to the visible black particulate matter
there is the fine material (less than 2.5 microns) that creates
large health problems.

- Smog is a mixture of air pollutants, both gases and particles,
that create a haze near the ground. Sulfate particles, created
when sulfur dioxide combines with other chemicals in the
air, and ozone are the main causes of smog and haze in most
of the United States.

- Ozone is a form of oxygen that contains three oxygen atoms
per molecule. (O_2, the form of oxygen that humans need to
survive, contains two oxygen atoms per molecule.) It is
common in Earth's atmosphere, where it blocks much of the
sun's ultraviolet radiation, preventing it from burning up
most forms of life. Though it is beneficial and necessary in
the atmosphere, ozone is also destructive and highly toxic to
humans. Ozone forms spontaneously from the energy of
sunlight in the air, but it can also form from other reactions,
such as sparks from electrical motors or the use of high

Aerial view of mountaintop
removal and reclamation in
the Indian Creek vicinity of
Boone County, West Virginia.
© *Library of Congress.*

Where Does Air Pollution Come From?

According to the Environmental Protection Agency (EPA), mobile sources, such as cars, trucks, buses, trains, airplanes, and boats, represent the largest contributor to U.S. air toxics. In 1999 as much as 95 percent of the carbon monoxide in typical U.S. cities came from mobile sources, according to EPA studies. More than half of all nitrogen oxide air pollution in the United States came from on road and non-road vehicles. The rest came from industry, such as power plants and factories. But the EPA states that the majority of all hydrocarbons (53 percent) and particulate matter (72 percent) comes from non-mobile sources such as power plants and factories.

voltage electrical equipment such as televisions. Fossil fuel pollution contributes nitrogen oxides and other organic gases that can react to create ozone. Ozone forms close to the ground on light sunny days, especially in cities.

- Sulfur dioxide is a by-product of burning fossil fuels. It is one of the key ingredients of acid rain. The United States Environmental Protection Agency (EPA) considers the reduction of sulfur dioxide emissions a crucial part of the effort to clean up the nation's air. The United States has set national air quality standards, and state and local governments are required to meet them.

- Nitrogen oxides are gases that contain nitrogen and oxygen in different amounts. Most of them are colorless and odorless. Almost all nitrogen oxides are created by the burning of fossil fuels in motor vehicles, power plants, and industry. Nitrogen oxides react with sulfur dioxide to produce acid rain. They also contribute to the formation of ozone near the ground, and they form particulate matter that clouds vision and toxic chemicals that are dangerous to humans and animals. In addition, they harm water quality by overloading water with nutrients. Finally, they are believed to contribute to global warming.

- Carbon monoxide is one of the main sources of indoor air pollutants. It forms from the burning of fossil fuels in appliances such as kerosene and gas space heaters, gas water heaters, gas stoves and fireplaces, leaking chimneys and

Accidental Death

Burning a charcoal grill or kerosene heater or running a car engine inside an enclosed space, such as a closed garage, can produce enough carbon monoxide to kill a person. Every year people die from inhaling concentrated carbon monoxide. Death comes easily and without warning because the victim often does not notice any symptoms; he or she simply gets sleepy from lack of oxygen, loses consciousness, and dies as carbon dioxide builds up in the blood.

furnaces, gasoline-powered generators, automobile exhaust in enclosed garages, and other sources. Carbon monoxide binds with the iron atoms in hemoglobin (the part of blood that carries oxygen) and prevents the blood from taking up enough oxygen to keep the brain running.

The United States and the individual states have passed various laws regulating air pollution. The Clean Air Act, passed in 1990, is one of the most important. It requires states to meet air quality standards, creates committees to handle pollution that crosses borders between states or from Mexico or Canada, and allows the EPA to enforce the law by fining polluters. It creates a program allowing polluting businesses to apply for and buy permits that let them release a certain amount of pollutants. Businesses can buy, sell, and trade these permits. They can receive credits if they release fewer emissions than they are allowed to produce.

One major difficulty with controlling air pollution is that some pollutants can travel thousands of miles from their sources. Certain types of air pollution in one state can originate from a coal-burning plant in another. For that reason, air pollution regulations must focus on large regions if they are to have any effect at all.

Acid rain

Acid rain is rain with small amounts of acid mixed into it. When sulfur dioxide and nitrogen oxides are released from burning fossil fuels, they mix with water and oxygen in the atmosphere and turn into acids. The acids in acid rain are not strong enough to dissolve a person, but they can contribute to environmental problems, such as the following:

A HEPA Filter?

Particulate matter is air pollution in the form of particles suspended in the air. Of special concern to human health are fine particles (of less than 2.5 microns) that are easily inhaled and can cause irritation to the eyes, nose, and throat, and may get into the lungs and either be absorbed by the bloodstream or stay embedded in the lungs to cause more serious breathing problems. Particulate matter has even been linked to an increased risk of heart attack in people with heart disease. Breathing in the particles may cause shortness of breath and chest tightness. A HEPA (HEP-ah) filter cleans particulate matter from indoor air when it is used in vacuum cleaners or air conditioning and heating units. A HEPA filter makes indoor air healthier because it is a "high efficiency particulate arresting" filter.

- Polluting lakes and streams, which can kill fish, other animals, and aquatic plants and disrupt entire ecosystems
- Damaging trees at high elevations
- Deteriorating the stone, brick, metal, and paint used in everything from buildings and bridges to outdoor artworks and historical sculptures
- Damaging the paint on cars
- Impairing visibility by filling the air with tiny particles
- Causing health problems in humans when the toxins in the rainfall go into the fruits, vegetables, and animals that people eat.

The EPA has an Acid Rain Program that limits the amount of sulfur dioxide that power plants can produce, and the program has reduced emissions somewhat. Reducing emissions overall should contribute to eliminating acid rain.

Global warming

Most scientists believe that the use of fossil fuels has changed the world's climate, and that this change is continuing. Burning fossil fuels releases gases called greenhouse gases, which include carbon dioxide, methane, and nitrous oxide. Greenhouse gases are good at trapping heat. When the sun's radiation hits Earth, some of the heat

Smog shrouds the skyline of the city of Los Angeles in a view from the Hollywood Hills. The city is famous for pollution. © *Andrew Holbrooke/Corbis.*

is reflected back into space. When greenhouse gases get into the atmosphere, they act like the walls of a greenhouse, holding the heat in so that it cannot escape back to space. Ordinarily, this would be a good thing, because life on Earth depends on keeping some of the sun's heat on the surface. Since the industrial revolution, however, the amount of greenhouse gases in the atmosphere has increased. The amount of carbon dioxide has increased 30 percent; the amount of methane has increased 100 percent; and the amount of nitrous oxide has risen 15 percent. These gases make the atmosphere better at keeping heat in. As a result, Earth's temperature has risen and continues to rise.

Kyoto Protocol

In 1997 many of the world's nations agreed to work together to reduce greenhouse gases and stop global warming. These nations signed an agreement in Kyoto, Japan, referred to as the Kyoto Protocol. The Kyoto Protocol sets targets for reducing emissions and deadlines for nations to meet those targets. The United States and Australia have not agreed to participate because the protocol does not place the same requirements on developing nations as it does on industrialized nations.

The increase in global temperatures can cause many problems. A possible effect is a rise in sea levels, which can change the shape of coastlines; cause changes in forests, crops, and water supplies; and harm the health of humans and animals. Fossil fuels account for 98 percent of carbon dioxide emissions, 24 percent of methane emissions, and 18 percent of nitrous oxide emissions.

Oil spills

When transporting petroleum, there is always the danger that the oil will leak out of its tank and contaminate the local environment. Many oil spills occur when a giant tanker ship crashes and the petroleum leaks out of the tank into the ocean. Spills can also happen when oil wells or pipelines break, or when tanker ships wash their giant tanks, rinsing the residue straight into the ocean.

When oil gets into the ocean, it quickly spreads over the surface of the water, forming an oil slick. The oil clumps into tar balls and an oil-water mixture called mousse. Seabirds and marine mammals get caught in the oil and die.

The 1989 wreck of the *Exxon Valdez* in Prince William Sound, Alaska, caused the worst oil spill that has so far occurred in North America. The ship hit and slid onto a coral reef. The accident allowed 38,800 tons of oil from the tanker to spread over 1,200 miles (1,930 kilometers) of shoreline, killing over one thousand sea otters and between 100,000 and 300,000 seabirds. At least 153 bald eagles also died from eating dead seabirds covered with oil. The cleanup cost nearly $3 billion, a large portion of that furnished by the United States government.

Almost 14,000 oil spills are reported each year in the United States. Usually, the owner of the oil or the tanker takes responsi-

Build a Better Tanker

Transporting oil safely is a big concern for the oil industry. Modern tankers are much stronger than older ones, and they are built with double hulls. Double hull means there are two layers of metal between the oil and the ocean. Double-hulled tankers are much less likely to be torn open if they run into rocks or coral reefs.

bility for cleanup. Occasionally, local, state, and federal agencies must help. The EPA takes care of spills in inland waters, and the United States Coast Guard responds to spills in coastal waters and deepwater ports.

The long-term effects of oil spills are not known. Though it appears that it is possible to clean up most of the oil and that the local ecosystem can recover, it also seems that some of the effects of oil are very long-lasting. The Prince William Sound environment still had some problems in the early 2000s: many animal species affected by the spill had still not recovered to their pre-spill numbers, and some oil remained on the region's beaches.

Economic impact of fossil fuels

Because they have been plentiful and are usually less expensive than other energy sources, fossil fuels supply nearly all of the world's energy. At the beginning of the twenty-first century the world economy is based on inexpensive fossil fuel. Almost all modes of transportation and industries require fossil fuels. Prices of consumer goods and services from food to airline tickets are partly determined by the cost of fuel. When the price of oil goes up, people who sell goods and services often must raise their prices because it costs more to make or deliver products.

As developing nations increase their use of automobiles, electricity, and other goods and services, their demands for fossil fuels increase. For example, oil consumption in China grew rapidly in the early twenty-first century. By 2003 China was consuming the second largest amount of oil in the world, behind the United States. China does not have sufficient fuel reserves to supply its own needs, so it must buy petroleum from other countries. Oil producers can raise their prices because they have several buyers competing to purchase their product.

Yet fossil fuels are still the cheapest source of power in the modern world. Alternative energy sources, such as solar power or hydrogen fuel cells, are much more expensive. Most people will not choose an expensive source of power when a cheap one is available, even if the cheap source contributes to pollution. For example, many coal-burning power plants still produce large amounts of pollution because the cost of controlling the pollution is deemed too expensive.

Societal impact of fossil fuels

Modern life would be impossible without fossil fuels, and in many ways fossil fuels have benefited people. The fact that fossil fuels are everywhere means that it is nearly impossible to take any action without using them. In many houses turning on a light uses fossil fuels. Shopping, eating, going to school, and sleeping in a heated or air conditioned home require the burning of fossil fuels. Fossil fuels are an important global issue. Countries have clashed over the issue of oil.

Air and water pollution are also global issues. The pollutants that come from fossil fuels can spread from country to country. Developing nations, such as Thailand and China, have been rapidly increasing the number of cars owned and of fossil fuel-powered factories and power plants, which has resulted in an increase in air pollution. International groups that want to protect the environment must balance air and water quality with the desire of poorer nations to improve their economies. The less developed countries feel that the countries of Europe and the United States were allowed to use fossil fuels to build their economies, regardless of the environmental consequences, and that they too should be given that opportunity without being forced to worry about pollution.

Issues, challenges, and obstacles in the use of fossil fuels

Fossil fuels are widely used and widely accepted. Nevertheless, there are ways to make fossil fuels less polluting, such as the use of clean coal technology and hybrid automobiles. These technologies have not yet become widespread, in part because they cost more than the methods that are currently used. As pollution increases and fossil fuels become harder to get, new methods of using fossil fuels will probably become more common.

PETROLEUM

Petroleum is the most widely used fossil fuel, supplying about 40 percent of the world's energy. Petroleum is also called oil. One

Is Petroleum Really a Fossil Fuel?

Some scientists in Russia and Ukraine believe that petroleum is not actually a fossil fuel but that it formed in Earth's crust from rocks and minerals rather than plants and animals. These scientists believe that the formation of oil requires higher pressure than the formation of coal and that there is not enough organic matter in Earth's deposits to explain the amount of petroleum available in large fields. Scientists in other countries disagree with this idea.

of the most important uses of petroleum is as fuel for motor vehicles. It can also be used to pave roads, to make other chemicals, and to moisturize skin.

Petroleum is a hydrocarbon, which means it is made up mostly of molecules that contain only carbon and hydrogen atoms. It also contains some oxygen, nitrogen, sulfur, and metal salts. The term *petroleum* encompasses several different kinds of liquid hydrocarbons. The main ones are oil, tar, and natural gas.

Origins of petroleum

The ingredients in petroleum include microscopic plants and bacteria that lived in the ocean millions of years ago. When they died, these plants and bacteria fell to the bottom of the ocean and mixed with the sand and mud there. This process continued for millions of years, and gradually the layers at the bottom were crushed by the layers above them. The mud became hotter, and the pressure and heat slowly transformed it. The minerals turned into a kind of stone called shale, or mudstone, and the organic matter turned into petroleum and natural gas.

Because they are not solid, petroleum and natural gas can move around. They seep into holes in undersea rocks such as limestone and sandstone, called reservoir rocks. These rocks are porous, meaning they have tiny holes in them that allow liquids and gases to pass through, and function as sponges. Because they are lighter than water, oil and gas migrate upward, although still trapped within Earth's crust. Sometimes the oil and gas end up in an area of rock that is not porous and is shaped in such a way that it can contain liquid and gas. This area becomes a reservoir, or geologic trap, that holds the petroleum and natural gas. Rock formations especially good at trapping hydrocarbons include anticlines, or layers of rock that bend downward; salt domes, or anticlines with a mass of rock salt at the core; and fault traps, or spaces between cracks in Earth's crust.

Within a trap, petroleum, natural gas, and water separate into layers, still within the porous reservoir rocks. Water is the heaviest and stays on the bottom. Petroleum sits on top of the water, and natural gas sits on top of the petroleum. Sometimes the natural gas and petroleum inside a trap find a path to the surface and seep out.

Finding petroleum

Geologists are scientists who study the history of Earth and its life as recorded in rocks. When looking for oil they want to find underground geologic traps because these traps often contain petroleum that can be removed by drilling. Geologists use a variety of techniques to find oil traps. They use seismology (syze-MAH-luh-jee), sending shock waves through the rock and examining the waves that bounce back. Geologists also study the surface of the land, examining the shape of the ground and the kinds of rocks and soil present. These scientists use gravity meters and magnetometers to find changes in Earth's gravity or magnetic fields that indicate the presence of flowing oil. They use electronic "sniffers" to search for the smell of hydrocarbons. Finding oil is difficult. Scientists searching for oil have only about a ten percent success rate.

Petroleum is present all over the world, but large concentrations of it exist in only a few places. These accumulations are called fields, and they are the places where oil companies drill for oil. The largest fields in the world are in the Middle East, especially in Saudi Arabia, Qatar, and Kuwait, and in North Africa. There are also large fields in Indonesia, Nigeria, Mexico, Venezuela, Kazakhstan, and several U.S. states, including Alaska, California, Louisiana, and Texas.

Extracting petroleum

Once an oil company finds oil in the ground, it has to get the oil out in order to sell it. First the company has to take care of legal matters, such as getting rights to the area it wants to drill. Once that is done, the company builds an oil well, or rig.

All oil rigs have the following basic elements:

- A derrick, which is a tall structure that supports the drill apparatus above ground

- A power source, such as a diesel engine, that powers electric generators
- A mechanical system, including a hoist and a turntable
- Drilling equipment, including drill pipe and drill bits
- Casing to line the drill hole and prevent it from collapsing
- A circulation system that pulls rock and mud out of the hole
- A system of valves to relieve pressure and prevent uncontrolled rushes of gas or oil to the surface

As oil workers drill deeper, they add sections of pipe to the drill and add casings to the hole to keep it stable. They drill until they reach the geologic trap that contains the oil and gas. To get the oil out of reservoir rocks, workers pump in acid or a fluid containing substances to break down the rock and allow the oil to seep into the well. The workers then remove the rig and install a pump in its place. The pump pulls the oil out of the well. Once the oil has been removed from the ground, the oil company must transport the

Some oil is located under the oceans. Oil drilling platforms are built on the water. These platforms are off the coast of Texas. © Jay Dickman/ Corbis.

If Petroleum Formed in the Ocean, Why are Oil Wells on Land?

When petroleum was forming, much of the area that is now dry land was covered with water. The ocean has moved away since then, but the oil is still there. In addition, many oil wells are out in the ocean, not on land at all.

crude oil to a refinery. The most common means of transporting oil are tanker ships, tanker trucks, and pipelines.

Making petroleum useful

Crude oil arrives at the refinery with a great deal of water and salt mixed into it. The water and oil are mixed together in droplets forming an emulsion, which is something like what happens to a salad dressing made of oil and vinegar. The water and oil may eventually separate out into their layers, but this process can take a very long time in thick crude oil. To speed the process, oil refineries heat the crude oil to a temperature at which the water can move more easily. The water molecules then come together and leave the oil. The water also takes the salt out of the oil with it.

The refinery distills crude oil to sort it into its different forms. Crude oil has many different kinds of molecules, some much larger than others. The refinery sorts out these molecules so that molecules of the same size are all together. A refinery is shaped like a tower with trays stacked one above the other. Heating the crude oil makes the molecules turn into gases. These gases move up inside the refinery's tower. As they travel upward in the tower, the gases become colder. At certain temperatures, they become liquids again. The liquids drip back down and are caught in one of the trays. The higher the gas travels, the higher the tray it ends up in. The largest molecules stay at the bottom. The smallest molecules make it all the way to the top of the tower. The lighter molecules are turned into gasoline and other fuels. The heavier ones become engine lubricants, asphalt, wax, and other substances.

There is a much larger market for gasoline and other fuels than for the products made from heavier molecules, so refineries try to make as much gasoline as possible. They can sometimes break down larger molecules into smaller ones. They do this through a process called cracking, which uses either heat or chemical catalysts to break down the large molecules.

The Oil Sands of Canada

Since the 1960s, investors and developers have been working to extract crude oil stored in the oil sands of Alberta, Canada. Some experts put the amount of proven oil reserves in the western Canadian oil sands at roughly 175 billion barrels. This would put it second only to Saudi Arabia (with 260 billion) in terms of proven oil reserves. Others believe that the amount of reserve oil in Alberta is much higher, possibly at 300 billion barrels, with more potentially buried deep underground.

Though people dreamed for decades of striking it rich by getting the oil out of Canada's sand, techniques are still in the early stages because of the difficulty of removing it. When compared to the relative ease of getting the oil that comes gushing out of oil fields in the Middle East and Texas, the existing process for turning oil sand into crude oil is difficult and expensive. It requires oversized trucks and shovels to dig out the sand and various machines to crush it, mix it with hot water, spin it to separate out the oil, and heat it to remove impurities. The expense concerned oil investors until political issues in the Middle East and other oil-producing nations and increasingly high demand in the early 2000s drove up oil prices to record levels, finally making oil removal from Alberta's sands profitable. With demand for crude oil on the world market growing, in particular to meet the needs of the United States and China, many of the residents and government officials of Alberta and Canada saw the potential for job creation and huge profits for the province and the rest of the country. In addition to making money by selling the oil, Canada could also potentially use the oil to negotiate with other countries on trade and political issues.

In the decades to come, Canada may become one of the biggest players in the fossil fuel economy, though the benefits may come at a high cost. The large amounts of natural gas and water used in the separation process create concerns for environmentalists. So does the excavation of thousands of tons of mud and sand, which creates large mining pits in Alberta's landscape. Though the oilmen who run Alberta's oil sand industry have promised to improve technology to clean up their greenhouse gas emissions and refill the mines and replant trees, groups like the Sierra Club of Canada have their doubts about whether technology will progress fast enough or trees grow quickly enough to make it worth the environmental damage. With little encouragement for conservation and the use of alternate energy sources by the worldwide community, demand for crude oil will most certainly transform Canada's economy and landscape as the oil sands become a valuable energy source for the world.

Current and potential uses of petroleum

Petroleum has many uses. It can take on different consistencies depending on how much it is refined. About 90 percent of the

The Isla Oil Refinery in Curacao, Netherlands Antilles. © 2005 Kelly A. Quin.

petroleum used in the United States is used as fuel for vehicles. Fuel types include:

- Motor gasoline used to power automobiles, light trucks, or pickup trucks that people drive as their daily transportation, boats, recreational vehicles, and farm equipment such as tractors
- Distillate fuel oil, including the diesel fuel used to power diesel engines in trucks, buses, trains, and some automobiles
- Heating oil to heat buildings and power industrial boilers
- LPGs (liquid petroleum gases), including propane and butane. Propane is used for heating and to power appliances. Butane is used as fuel and is blended with gasoline
- Jet fuel, which is a kerosene-based fuel that ignites at a higher temperature and freezes at a lower temperature than gasoline, making it safer to use in commercial airplanes
- Residual fuel oil used by utilities to generate electricity
- Kerosene used to heat homes and businesses and to light lamps

Stopping the Knocking

The question of how to prevent engine knocking has occupied petroleum engineers for many years. In the mid-twentieth century, they added lead to gasoline to make it burn more efficiently. In 1979 leaded gasoline became illegal in the United States due to fears of lead poisoning in children. Since that time MTBE (methyl tertiary-butyl ether) has been added to gasoline in the United States to enhance octane. It has done a great job of reducing emissions from car engines, but it is not perfect. People are concerned that MTBE is dangerous when it gets into drinking water, and they want to find a substitute. Ethanol has been used in some cases, but it has drawbacks, too. As of the early 2000s, oil companies were still looking for the perfect fuel additive.

- Aviation gasoline, which is a high-octane gasoline used to fuel some aircraft
- Petroleum coke used as a low-ash solid fuel for power plants and industry

Petroleum has many other uses, including:

- Petrolatum, or petroleum jelly, used as a moisturizer and lubricant
- Paraffin wax used in candles, candy making, matches, polishes, and packaging
- Asphalt or tar used to pave roads or make roofs
- Solvents used in paints and inks
- Lubricating oils for engines and machines
- Petroleum feedstock used to make plastics, synthetic rubber, and chemicals

The United States uses over 250 billion gallons of oil every year. About one-half of that amount comes from domestic wells; the other half is imported.

Benefits and drawbacks of petroleum

As compared to other fossil fuels, petroleum is easy to retrieve, refine, and use. It is fairly easy to transport and store. It is not prone to exploding spontaneously, so it is relatively safe to keep near homes. Petroleum burns easily, making it the ideal fuel for internal combustion engines. Petroleum has many applications in addition to fueling vehicles. These uses range from paving materials to skin moisturizers.

Using petroleum, however, has many drawbacks. It contributes to various types of environmental problems, including air and water pollution. There is only a limited supply of petroleum, which means that at some time in the future, the world's petroleum will be gone. When that happens, people will have to find another way of powering their vehicles, factories, and utilities.

Impact of petroleum

Using petroleum as fuel contributes to many environmental problems. These include oil spills, which typically happen during the transportation of petroleum; the destruction done by drilling for oil; contamination from oil wells and pipelines; and air pollution. Drilling for oil, for instance, requires massive pieces of equipment and results in giant holes in the ground. Contamination happens when oil seeps into local soil and water. The people who live near oil wells and refineries sometimes suffer health problems as a result of exposure to petroleum.

When gasoline burns, it releases carbon dioxide and water into the atmosphere. It also produces carbon monoxide, nitrogen oxides, and unburned hydrocarbons, all of which can contribute to air pollution. Modern automobiles use catalytic converters to remove some of the pollutants from car emissions. Because of this improvement in car technology, automobiles in 2005 produced much less pollution than cars in 1970.

The economic impact of petroleum is enormous. The United States uses more than seventeen million barrels of oil daily and 250 billion gallons of oil a year. More than one-third of that petroleum powers cars and trucks. The country must import more than one-half of that amount from other countries. The United States has more oil reserves than it currently uses, but as of the early twenty-first century it was much less expensive to import oil than it was to extract reserves within the country. Foreign oil is becoming more expensive, however, especially as other countries increase their oil consumption. Some people support opening new U.S. sites to oil exploration and drilling partly because the oil industry can create so many good jobs.

A sudden change in oil prices can be disruptive to the United States and world economies. For example, oil prices rose steeply in

the 1970s, creating an oil shock and inspiring car manufacturers to improve fuel economy. Oil prices were low and stable during the late 1980s and most of the 1990s. Rising prices in the early 2000s reflected increased demand for oil and other complicated economic factors.

The Zueitina Oil Company's excess oil, water, and product waste dumping ground is outside the main oil pumping facilties in Libya. © *Benjamin Lowy/Corbis.*

Issues, challenges, and obstacles in the use of petroleum

There is a limited supply of petroleum on Earth. Some experts believe that oil production will peak by 2020 and that current oil reserves will run out by 2050, if not earlier. Other experts disagree, believing that there are enough oil reserves to provide for the world's energy needs throughout the twenty-first century. Many areas in the Middle East and Russia are still unexplored. Oil companies can now drill in much deeper parts of the ocean than they previously could; oil rigs in the Gulf of Mexico now drill into wells below 10,000 feet (3,048 meters) of water. Improved drilling technology such as drills that can twist and turn underground allows oil companies to reach petroleum deposits miles away from rigs.

OPEC

The Middle East holds a great deal of the world's petroleum. Middle Eastern nations and a few others have formed an organization called the Organization of the Petroleum Exporting Countries, or OPEC (OH-peck), which coordinates the prices that the individual countries charge for oil. Most OPEC member countries are developing nations, which means they are working on making their countries more modern. Oil is extremely important for these countries because it brings in a huge amount of money. Furthermore, as long as petroleum is needed, the OPEC nations will have power over the rest of the world.

Pessimists argue that improved technology will only deplete oil reserves faster, especially as more of the world uses oil to power its vehicles and industry. Optimists believe that should not matter and that innovation will allow oil companies to keep furnishing the world with petroleum.

NATURAL GAS

Along with coal and petroleum, natural gas is one of the three main fossil fuels in use in the early twenty-first century. People use natural gas for heating, electrical power, and other purposes. Natural gas produces much less pollution than petroleum, so some people believe it could be an ideal substitute for petroleum and coal in the future.

Natural gas is a gaseous hydrocarbon. It is colorless, odorless, and lighter than air. Natural gas is made up of 75 percent methane, 15 percent ethane, and small amounts of other hydrocarbons such as propane and butane.

The substance that oil companies sell as natural gas is almost pure methane, with the other gaseous components removed. When it burns, methane releases a large amount of energy, which makes it a useful fuel. Methane is sometimes called marsh gas because it forms in swamps as plants and animals decay underwater. Methane is naturally odorless, but gas companies add traces of smelly compounds to natural gas so that people will be able to smell gas leaks and avoid danger.

Oil wells at Midway-Sunset
Oil Field in California.
© *Lowell Georgia/Corbis.*

Origins of natural gas

Natural gas formed from underwater plants and bacteria. These microscopic organisms fell to the bottom of the ocean when they died and over the course of millions of years were crushed and heated by the pressure of layers of sand, dirt, and other organic matter that accumulated on top of them. The mineral components of the undersea mud gradually turned into shale, and some of the

Fuel Economy in Cars

The average fuel economy of cars sold in the United States has decreased steadily since 1985. That means that on average, a new car today uses more fuel for the same performance than an equivalent car built 20 years ago. Sport utility vehicles (SUVs) are partly to blame for this. The Clean Air Act required car manufacturers to build vehicles to certain specifications that limited pollution. Certain types of vehicles, such as trucks, did not have to meet the same standards as cars because they were larger and there were relatively few people driving them at the time. Because SUVs are classified as light trucks under the law, they do not have to have the same level of fuel economy that a passenger car has. Car manufacturers like SUVs because they are inexpensive to make and can be sold for relatively high prices. SUVs have also been fashionable among consumers. In 2005 a proposed reform of the government's Corporate Average Fuel Economy (CAFE) program for light trucks by the National Highway Traffic Safety Administration (NHTSA) required carmakers to make gradual changes to their designs to meet stricter fuel economy requirements for light trucks by 2011. The proposed plan was scheduled to go into effect in April 2006.

As gas prices rose and concerns about America's dependence on foreign oil began to concern Americans in the early 2000s, hybrid cars became fashionable. These cars were powered by a combination of gasoline and battery power and had considerably better mileage than gasoline-powered cars. In their first years only a few were available so they were hard to buy. Some critics complained that hybrid cars were too expensive and that they did not in fact provide the fuel economy that a small, light, efficient gasoline-powered car could.

organic components turned into natural gas. Natural gas can move around within porous reservoir rocks. It can also be trapped in underground reservoirs, or geologic traps. Natural gas is lighter than petroleum, so it usually sits on top of the petroleum in a reservoir. Natural gas sometimes seeps up through Earth's crust and appears on the surface.

Finding and extracting natural gas

Natural gas is usually found with petroleum. When geologists (scientists who deal with the history of Earth and its life as recorded in rocks) search for underground oil, they find natural gas along with it. Sometimes there are pockets of natural gas in coal beds. Geologists occasionally find reservoirs that contain mostly or all natural gas with no oil. The largest reserves of natural gas in the United States are in Texas, Alaska, Oklahoma, Ohio, and

Make Your Own Methane

Although most of the world's methane is very old, it is possible to make new methane through chemical reactions. The Sabatier process combines hydrogen and carbon dioxide with a nickel catalyst and high temperatures to synthesize methane and water. This method of producing methane could be used to generate fuel in outer space to power spacecraft. One common natural process also results in large amounts of methane: When cattle digest food, they produce methane that they emit into the atmosphere.

Pennsylvania. Some experts believe that there is enough natural gas in the Earth to last two hundred years, although much of this gas may be difficult to reach.

When they first began drilling for oil, people believed natural gas was an unpleasant by-product. They would burn the natural gas away before removing the oil from the ground. Now oil companies know that natural gas is a valuable commodity in its own right, and they extract it carefully. The process of drilling for natural gas is similar to that of drilling for petroleum. In many cases natural gas comes out of wells that have already been dug to extract oil. Oil companies also drill wells to extract natural gas by itself. There are three main kinds of natural gas wells:

- Gas wells, which are dug into a reservoir of relatively pure natural gas
- Oil wells, which are dug for extracting oil but also extract any natural gas that happens to be in the reservoir
- Condensate wells, which are dug into reservoirs that contain natural gas and a liquid hydrocarbon mixture called condensate but contain no crude oil

Natural gas that comes from oil wells is sometimes called associated gas. Natural gas from gas wells and condensate wells is called non-associated gas because it is extracted on its own and not as a by-product of oil drilling.

Making natural gas useful

The natural gas that consumers use is almost pure methane. The natural gas that comes out of a well is not pure and may contain a mixture of hydrocarbons and gases, including methane, ethane, propane, and butane. It also may contain small amounts of oxygen, argon, and carbon dioxide, but methane is by far the largest component.

An oil or gas company processing natural gas separates the gases into individual components, dividing them into pure methane, pure propane, pure butane, and so on. The liquid forms of the non-methane gas components, such as propane and butane, are called natural gas liquids, or NGLs, and sometimes are called liquid petroleum gas, or LPG. All of these products can be sold individually, so it is cost-effective to separate them.

The first step in processing is to remove any oil mixed with the gas. Natural gas that comes out of an oil well is separated from petroleum at the well. Sometimes the gas is dissolved in the oil, like the carbonation in a soft drink, and through the force of gravity the gas bubbles come out of the oil. In other cases the oil workers use a separator that applies heat and pressure to the mixed oil and gas to make them separate. The workers must also remove any water from the natural gas, using heat, pressure, or chemicals. They then remove NGLs using similar techniques.

Once they have been removed from natural gas, NGLs must be separated from one another. This is done through a process called fractionation, which involves boiling the NGLs until each one has evaporated. A similar process is used to refine petroleum. The different NGLs have different boiling points. As the NGLs boil, the different hydrocarbons evaporate and can be captured.

Some natural gas comes out of the ground with large amounts of sulfur in it. It is called sour gas because the sulfur makes the gas smell like rotten eggs. The gas company must remove the sulfur before selling the gas because sulfur in significant amounts is poisonous for humans to breathe and because it corrodes metal. The companies can sell the sulfur for industrial uses once it is separated out.

Sometimes a processing plant turns natural gas into liquid before transporting it. Liquid natural gas is one six-hundredth the volume of natural gas in gas form. Liquefying it makes it possible to store and transport natural gas around the world.

Once it has been refined and liquefied, natural gas can be transported and sold. The most common way to transport natural gas is through pipelines, which crisscross the United States and many other countries. If the gas is not sold right away, the gas company must store it. Natural gas is usually stored underground in formations such as empty gas reservoirs; in aquifers, or underground rock formations that hold water; and in salt caverns.

Current and potential uses of natural gas

People have known about natural gas for thousands of years. The eternal flames in ancient temples may have been fueled by natural gas. In the early nineteenth century people began using natural gas as a light source, but as soon as oil was discovered in the 1860s and electricity became widespread, people abandoned natural gas except for limited use in cooking and heating.

Even so, the natural gas industry built the first large natural gas pipeline in 1891 and a large network of pipelines in the 1920s. Gas companies built more pipelines between 1945 and 1970, which made it convenient to use natural gas for heating homes and for use in appliances.

Natural gas has become more appealing as a fuel in recent years. Some uses are:

- Powering heaters and air conditioners. Because so many homes and businesses use gas heat, natural gas consumption typically is much higher in the winter than in the summer.
- Running appliances such as water heaters, stoves, washers and dryers, fireplaces, and outdoor lights.
- Serving as an ingredient in plastics, fertilizer, antifreeze, and fabric.
- Producing methanol, butane, ethane, and propane, which can be used in industry and as fuel.
- Dehumidifying, or drying the air in, factories that make products that can be damaged by moisture.

Scientists are considering the use of natural gas in applications such as the following:

- Powering natural gas-fueled vehicles, which produce far fewer emissions than vehicles powered by gasoline.
- Powering fuel cells in which hydrogen is used to produce electricity with few emissions.
- Reburning, or adding natural gas to coal- or oil-fired boilers to reduce the emission of greenhouse gases.
- Cogeneration, a technology for generating electricity as it burns fuel, requiring less total fuel and producing fewer emissions.

- Combined cycle generation, a technology that captures the heat generated in producing electricity and uses it to create more electricity. Combined cycle generation units powered by natural gas are much more efficient than those powered by petroleum or coal.

Scientists are especially interested in technologies that combine natural gas with other fossil fuels to increase efficiency and reduce emissions. Natural gas is seen as a good source of fuel for the future, and as a result scientists are constantly inventing new ways to use it.

Benefits of natural gas

Natural gas has advantages over petroleum and coal. It burns cleanly, producing no by-products except for carbon dioxide and water, so it does not cause the same degree of air pollution as the other fossil fuels. It does not produce the sludge that results from coal-burning emissions.

Natural gas can take the place of gasoline as a fuel for cars, trucks, and buses. Most natural gas vehicles are powered by compressed natural gas (CNG); the technology used to pump CNG into a car is almost identical to the process of fueling a gasoline-powered car. Some vehicles can use either gasoline or CNG. Natural gas cars have no trouble meeting environmental standards because of their low emissions. Natural gas is very safe; it does not pollute groundwater.

For many years natural gas has been cheaper than gasoline. Many cities have converted their buses, taxis, construction vehicles, garbage trucks, and public works vehicles to natural gas. These organizations are well suited to use natural gas as fuel because their vehicles do not travel long distances and can afford the cost of converting the vehicles in the first place.

Drawbacks of natural gas

Natural gas historically was hard to transport and store, but modern technology has for the most part removed that difficulty. One reason natural gas is not a perfect substitute for petroleum is that supplies are limited. At current rates of use, all of the world's natural gas could be used up in forty to ninety years.

Natural gas vehicles have not become widespread because it is more expensive to convert gasoline vehicles for natural gas use; there are very few natural gas refueling stations; and the vehicles cannot travel long distances without refueling.

Impact of natural gas

Natural gas is the cleanest fossil fuel. The burning of natural gas releases no ash and produces low levels of carbon dioxide, carbon monoxide, and other hydrocarbons and very small amounts of sulfur dioxide and nitrogen oxides. Vehicles powered by natural gas emit 90 percent less carbon monoxide and 25 percent less carbon dioxide than gasoline-powered vehicles.

Natural gas is becoming an increasingly common fuel for electrical power plants and in industry. Electrical power plants fueled by natural gas produce far fewer emissions than coal-powered plants. Burning natural gas does not contribute significantly to the formation of smog.

Natural gas does contribute to some environmental problems. Burning natural gas emits carbon dioxide, which is considered a greenhouse gas that contributes to global warming. On the other hand, natural gas produces 30 percent less carbon dioxide than burning petroleum and 45 percent less carbon dioxide than burning coal, so it is still preferable to either of those.

On an economic level, the cost of natural gas has dropped considerably. The development of LNG technology means that natural gas is easier and less expensive to store and to transport, and liquefaction techniques (turning gas into a liquid) improve every year. Petroleum engineers are constantly getting better at finding and extracting natural gas from the ground.

Natural gas may change the way people use power in their daily lives. In the twenty-first century natural gas is a fairly minor fuel compared with gasoline, but it has the potential to be much more important. If power plants switch to the use of natural gas during summer when demand for natural gas is lowest and smog is highest, they could emit fewer pollutants and improve air quality. Using natural gas instead of other fossil fuels could reduce acid rain and particulate emissions. As people become concerned about emissions and fuel economy, they may want vehicles powered by natural gas. The vehicles will then become more widely available, less expensive, and easier to refuel.

Issues, challenges, and obstacles in the use of natural gas

Natural gas technology is not widespread. The fuel has many possible applications, but car manufacturers will have to decide that it is cost-effective for them to build natural gas vehicles before they do so on a large scale. Consumers will not buy natural gas vehicles until they are convinced that it will be convenient, safe, and inexpensive for them to buy natural gas as fuel. A final large issue is the supply of natural gas, which could run out in a few decades.

COAL

Coal supplies about one-fourth of the world's energy needs. Coal is a solid hydrocarbon made primarily of carbon and hydrogen with small amounts of other elements such as sulfur and nitrogen. Coal looks like black rock, and it leaves black dust on things that it touches.

Origins of coal

Millions of years ago Earth was covered with swamps full of giant trees and other plants. When they died, these trees fell into the swampy water and were gradually covered by other plants and soil. All living things, including plants and animals, are composed mainly of carbon. Over millions of years, the carbon in the swamp plants was compressed and heated. This caused it to rot, exactly the way fruit and vegetables rot if kept too long. This rotting produced methane gas, also known as swamp gas.

Over several thousand years, the weight of the upper layers compacted the lower layers into a substance called peat. Peat is the first step on the way to the formation of coal and other fuels. People can use peat as fuel simply by cutting chunks of it out of the ground and burning them. Ireland used to be covered with peat, which was the main source of fuel there for years. The Great Dismal Swamp in North Carolina and Virginia contains almost one billion tons of peat.

As the peat continued to be compacted by new layers of dead plants, it became hotter as it was being pushed closer to the heat inside the Earth. The heat and pressure gradually turned it into coal. Most of Earth's coal was formed during one of two periods: the Carboniferous (360 million–290 million years ago) or the Tertiary (65 million–1.6 million years ago).

Finding coal

There are large reserves of coal all over the world. China has nearly one-half of the world's coal reserves and produces nearly one-fourth of the coal that is used every year. There are also large reserves of coal in North America, India, and central Asia. In the United States, most coal comes from mines in Montana, North Dakota, Wyoming, Alaska, Illinois, and Colorado. There are also coal deposits in the Appalachian area, especially in West Virginia and Pennsylvania.

Getting coal out of the ground

Coal is extracted from the earth through mining techniques that vary depending on where the coal is located. If a coal seam (or deposit) is deep below the surface of the Earth, miners use subsurface mining. They dig vertical tunnels into the ground to reach the seam and then dig horizontal tunnels at the level of the seam. The miners ride elevators down to the seam, dig out the coal, and transport it back up to the surface. To prevent the earth from collapsing, miners leave pillars of coal standing to hold up the tunnel roof. Despite this precaution, coal mines sometimes collapse, killing miners trapped inside.

Surface mining, or strip mining, is a process of taking coal off the surface of the Earth without going underground. Miners use giant shovels to remove dirt, called overburden, from the coal seam and then use explosives to blast the coal out of the rock. Strip mining is much safer than subsurface mining, but it leaves huge scars on the land and can contribute to water pollution.

Making coal useful

Coal comes out of the ground in chunks up to 3 or 4 feet (0.9–1.2 meters) across, and coal processors crush it into chunks about the size of a person's fist. These chunks of coal then go through a screen that separates out the smallest pieces. Coal plants sometimes clean coal by setting it, which washes out the heavier particles of stone. The plant may then dry the coal to make it lighter and help it burn better. Once processing is complete, coal is transported to buyers using trains, barges (flat cargo-carrying boats), and trucks.

Coal comes in several types, depending on how pure the carbon is, which also corresponds to how old the coal is. Coal is rated by heat value (how much heat it can produce when it burns). The purer the carbon is, the higher the heat value. Heat value is measured in British thermal units, or Btu, per pound. A Btu is the amount of heat required to raise the temperature of one pound of water one degree Fahrenheit.

- Anthracite (AN-thruh-syte) contains between 86 and 98 percent pure carbon and has a heat value of 13,500 to 15,600 Btu per pound.
- Bituminous (bye-TOO-muh-nuhs) coal contains between 60 and 86 percent pure carbon and has a heat value of 8,300 to 13,500 Btu per pound.
- Lignite contains between 46 and 60 percent pure carbon and has a heat value of 5,500 to 8,300 Btu per pound.

Current and potential uses of coal

Coal became a popular fuel in England in the nineteenth century because England sits on top of huge coal deposits. Coal was more plentiful than wood, which meant it was less expensive. The availability of coal along with inventions such as the steam engine allowed England to become the first truly industrialized nation.

During the nineteenth century and in the early part of the twentieth century, many people had coal-burning stoves in their homes. This system of heating had many drawbacks. It was messy, and people had to make sure they did not run out of coal. By the late twentieth century coal was no longer a common fuel for heating homes. As individual homeowners used less coal, industry used more.

Between 1940 and 1980 the amount of coal used by electrical power plants doubled every year. Coal also powers factories that make paper, iron, steel, ceramics, and cement. At the beginning of the twenty-first century over one-half of the electrical power plants in the United States were powered by coal.

Benefits and drawbacks of coal

Coal burns hotter and more efficiently than wood, and in many places it is more readily available. There is a great deal of coal in the world, so supplies are not likely to run out in the near future.

One of the drawbacks of using coal is that it has to be dug out. All methods of mining coal have problems associated with them. Coal is also very dirty. Coal dust coats anything it falls on, from buildings to people. Gases released by burning coal are big contributors to air pollution.

Environmental impact of coal

Coal is not environmentally friendly. It produces large amounts of pollution, which may contribute to acid rain and global warming. Mining it is often damaging to the environment, and transporting it is destructive as well. Most coal is moved around on trains, which are powered by pollution-causing diesel fuel.

Air pollution

The difficulty with burning coal is that it rarely produces only carbon dioxide, water, and energy. If the temperature is not high enough or if not enough oxygen is available to keep the fire burning high, the coal is not completely burned. When that hap-

Eternal Coal Fires

Sometimes the coal inside a mine will catch on fire by accident. It can be nearly impossible to put out this kind of fire; drilling into the mine only adds oxygen to fuel the flames. A coal deposit in Tajikistan has supposedly been burning underground since 330 BCE when Alexander the Great visited the area. A network of coal mines in Centralia, Pennsylvania, caught fire in 1962 and is still burning. Someone had burned trash in an abandoned coal pit, and the coal vein ignited. The town had to be evacuated in the 1980s. Hundreds of coal mines are burning in the United States, but many more are burning in China and India, where mining development is proceeding too rapidly to control. In addition, coal mining produces tailings (coal mining wastes) that are put in large piles above ground; the tailings can also catch fire and burn for decades.

pens, the coal releases other substances into the air. These substances include:

- Carbon monoxide, which is toxic to humans and animals
- Soot, which is pure carbon dust and can turn buildings, trees, and animals black (The English invented glass-covered bookcases in the 1800s so their books would not get covered with soot.)
- Sulfur dioxide, sulfur trioxide, and nitrogen oxides, which become part of acid rain
- Lead, arsenic, barium, and other dangerous compounds that are in coal ash, which can float in the air or stay where the coal was burned and cause people to become ill

As mentioned, electrical power plants produce 67 percent of the United States sulfur dioxide emissions, 40 percent of carbon dioxide emissions, 25 percent of nitrogen oxide emissions, and 34 percent of mercury emissions. Coal-fired power plants account for over 95 percent of all these emissions.

New power plants may be less polluting than older ones, but most power plants operating in the United States as of 2005 still used older technology. Under the Clean Air Act older plants were prevented from expanding, in the hope that they would gradually close down and be replaced by modern facilities. The Clear Skies program enacted in 2003 by President George W. Bush removed this requirement, allowing older plants to keep operating and to expand their operations if they chose to do so.

Regardless of what developed countries of the twenty-first century do about emissions, China and other developing nations are using outdated technology that releases huge amounts of pollution. As the developing nations move towards resembling the developed world technologically, vast amounts of pollution travel around the world and end up in countries elsewhere.

Coal mining

Surface coal mining can leave huge holes in the land and even destroy entire mountains. Water that flows over the mine site can flush pollutants into streams and rivers. Underground coal mining leaves behind tunnels in the ground, which can collapse suddenly. In the old days of mining, abandoned surface mines would turn into forbidding deserts, full of old rusted equipment.

Modern coal mining is very different, at least in the industrialized world. Due to several decades of pressure from consumers and environmental groups and new environmental laws, twenty-first century coal mining companies are much more careful about restoring the landscape after they take the coal from it. Miners save the topsoil and store local plants in greenhouses. Mining companies hire biologists, botanists (scientists who study plants), and fisheries experts to restore the environment as it was before mining began. Before laws required it, no mining company spent the money to avoid environmental harm.

Economic impact of coal

Coal started the industrial revolution in Europe in the late eighteenth century. Without coal, there would have been no factories, no steel, no trains, no steamships, and no electric lights. In the early twenty-first century coal is still a huge business. Coal mines bring in a great deal of money. In areas that have large coal deposits, most of the local population may be employed by the coal industry. The closing of a coal mine can harm a community by putting many townspeople out of work.

Societal impact of coal

Coal mining was one of the first industries to attract the attention of socially conscious lawmakers, who passed laws protecting workers. Coal mining was also one of the first industries in which workers organized, leading to the development of trade unions. Although mining techniques in the United States are much better than they were in the nineteenth century, coal miners still face more daily risks than most workers. Some health problems are much more common in coal miners than in other groups of people. Aside from the danger of being killed in a mine collapse, coal miners are at risk of life-threatening lung diseases. People who live in coal mining regions depend on the coal industry for their income and do not want to see coal mining disappear. At the same time, they would like to see coal mining become safer and less destructive.

Issues, challenges, and obstacles in the use of coal

The demand for coal is expected to triple in the twenty-first century. Coal is the only fossil fuel that is likely to be in large supply in the year 2100, so people may become even more dependent on it. The U.S. Congress has encouraged coal producers to clean up coal technology since 1970. Scientists are trying to invent ways to use coal for fuel without causing pollution. These methods are called clean coal technologies and include the following:

- Coal gasification, by which coal is turned into gas that can be used for fuel, leaving the dangerous solid components in the mine

- Coal liquefaction, by which coal is turned into a petroleum-like liquid that can be used to power motor vehicles

- Coal pulverization, by which coal is broken into tiny particles before it is burned

- Use of hydrosizers, which are machines that use water to extract (take out or remove) the usable coal from mining waste to increase the amount of coal that can be retrieved from a mine

- Use of scrubbers and other devices to clean coal before, during, and after combustion to reduce the amount of pollution released into the atmosphere

- Use of bacteria to separate pollutants from organic components in coal so that the sulfur and other pollutants can be removed before burning

- Fluidized bed technology, which burns coal at a lower temperature or adds elements to the furnaces in coal plants to remove pollutants before they burn

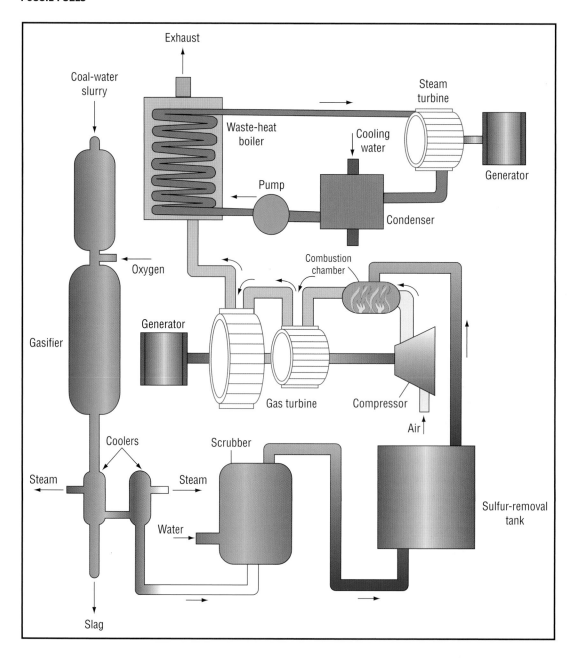

Exhaust

Coal-water
slurry

Steam
turbine

Waste-heat
boiler

Cooling
water

Generator

Pump

Condenser

Oxygen

Combustion
chamber

Gasifier

Generator

Gas turbine

Compressor

Air

Coolers

Scrubber

Steam

Steam

Water

Sulfur-removal
tank

Slag

Illustration showing coal gasification, with the elements that are used to give power to the electric generators. © *Thomson Gale.*

COAL GASIFICATION

Coal gasification is a process that converts coal to a gas that can be used as fuel. The main advantage of gasification is that it can remove pollutants from coal before the coal is burned, so the harmful substances are not released into the air. Coal gasification is a clean coal technology.

Coal gasification is done in stages. The first step is to crush and dry the coal. The crushed coal is placed in a boiler, where it is heated with air and steam. This heat causes chemical reactions that release a mix of gases that can then be used as fuel. The solid waste, or ash, remains in the boiler, where it can be collected and thrown away. Dangerous gases such as carbon dioxide and sulfur dioxide are removed in scrubbers like the ones in smokestacks at coal plants.

Gasification has been around for at least 100 years. It was widely piped and used as a fuel in Britain and many other European countries by 1900. Although it was used in other countries, in the United States it wasn't utilized during the first half of the century because petroleum and natural gas were inexpensive and plentiful. In the 1970s utility companies began considering gasification as a way to obey stricter environmental laws. Many people hope that coal gasification will be a valuable technology in the twenty-first century.

Current and potential uses of coal gasification

Coal gasification produces the following kinds of gases that can be used as fuel:

- Methane, which can be used as a substitute for natural gas
- Chemical synthesis gas consisting of carbon monoxide and hydrogen, which is used in the chemical industry to produce other chemicals, such as ammonia and methyl alcohol
- Medium-Btu gas, which is also made of carbon monoxide and hydrogen and used by utilities and industrial plants

Benefits and drawbacks of coal gasification

Plants and factories that run on coal gasification technology have much lower emissions than traditional coal-burning plants, and their solid wastes are not hazardous. The waste products themselves can be useful. The sulfur dioxide scrubbers produce pure sulfur that can be used in other processes, and some scientists believe the ash can be used to build roads and buildings. Some people believe it may even be possible to use sewage or hazardous wastes to power the coal gasification boilers.

The greatest problem with coal gasification is cost. Using coal gasification technology to provide power to an industrial plant costs three times as much as using natural gas. Supporters of the technology hope that researchers will develop ways to make gasification less expensive. Coal gasification requires vast amounts of water, which creates a problem. For gasification to be cost-effective, the plants must be built near coal mines so that the coal does not have to travel far, and most coal mines in the United States are in western states, where water is limited and expensive.

Impact of coal gasification

On an environmental level, gasification has the potential to make coal a much less polluting fossil fuel. It will not have any impact on the environmental destruction caused by coal mining itself. However, coal mining is now much less destructive than it used to be.

Economically, coal gasification is much less efficient than burning coal directly; 30 to 40 percent of coal's energy is lost during the process of converting it to gas. Gasification would hardly be worth the cost of production if it were not for the environmental benefits it offers.

Issues, challenges, and obstacles in the use of coal gasification

Scientists in Europe and the United States have been working to improve coal gasification techniques. They have been experimenting with using chemicals called catalysts to release the gases from coal. Using catalysts would allow gasification to occur at a lower temperature, which would make the process less expensive. Some scientists believe that the answer is to carry out gasification inside coal mines. Miners could pipe up the useful gases and leave the solid wastes underground. This idea is attractive because a large portion of coal reserves are nearly impossible to remove by the usual methods, and underground gasification would make those reserves available.

LIQUEFIED PETROLEUM GAS: PROPANE AND BUTANE

Liquefied petroleum gas, or LPG, is petroleum gas that can easily be turned into a liquid at ordinary temperatures simply through the application of pressure. The main types of LPG are propane and butane. Propane is the most common LPG and is usually what people mean when they refer to LPGs. Propane and butane are both colorless, flammable gases that belong to the category of hydrocarbons called paraffins or alkanes. Unprocessed natural gas contains both propane and butane, which are removed during the purifying process. Petroleum refining also creates LPGs.

The first step in processing LPG is to remove any oil that might be mixed with the gas. Sometimes the natural gas is dissolved in oil, and the gas bubbles will come out of the oil through the force of gravity. In other cases oil workers use a separator that applies heat and pressure to the mixed oil and gas to make them separate.

Once the methane has been removed from the natural gas, the workers separate the remaining components, which include propane, butane, and ethane in a liquid form. The process is called fractionation, which basically involves boiling until each one of the gases has evaporated. The different gases have different boiling points. As each different boiling point is reached, the gases evaporate and can be captured separately. Because LPGs are naturally odorless, oil companies often add a substance called ethanethiol (eth-THAN-ee-thee-all) to it so people can smell the gas if it leaks. Ethanethiol smells like rotten eggs.

Oil companies usually store large amounts of LPGs in underground salt domes and pressurized empty mines near gas production facilities and pipeline hubs. These reservoirs are tied directly to pipelines so the LPGs can be delivered rapidly. LPG merchants store the gas in large pressurized above-ground tanks. Consumers then store LPGs in smaller above-ground tanks at their homes or businesses.

Most LPGs in the United States are transported through a network of about 70,000 miles (113,000 kilometers) of pipelines. Most of these pipelines are concentrated along the Gulf Coast and in the Midwest. The Midwest also receives LPGs from two pipelines running from Canada. The east coast of the United States has only two pipelines serving the area. LPGs can be delivered by trucks, trains, barges, and ocean tankers. The United States imports about ten percent of its total LPG supply from other countries, including Saudi Arabia, Algeria, Venezuela, Norway, and the United Kingdom.

Current and potential uses of LPG

LPGs are useful as substitutes for natural gas for purposes such as powering stoves, furnaces, and water heaters. LPGs, often sold as or called propane, can be used in many ways, including:

- As a fuel for internal combustion engines, such as the ones in cars and buses

- To power home appliances, such as hot water heaters, heat pumps, space heaters, fireplaces, stoves, and clothes dryers
- As a fuel for devices such as forklifts
- For industrial purposes such as soldering, cutting, heat treating, and space heating
- To power campers and recreational vehicles
- As a solvent and refrigerant in the petroleum industry
- As a propellant in aerosol sprays, replacing CFCs (chlorofluorocarbons)
- For agricultural purposes such as weed control, crop drying, and as fuel for irrigation pumps and farm equipment

Butane by itself is used in cigarette lighters and portable stoves, such as the stoves people take camping. Petroleum refineries leave some butane in gasoline to make it easier to start engines since butane ignites quickly.

Ethane, which is another kind of LPG, is used as a starting material in the production of ethylene and acetylene, which are used as fuel in welding. It is possible to power automobiles and other vehicles with LPG. Some people have converted their cars to burn LPG instead of gasoline.

Homeowners and private consumers use about 45 percent of the LPGs sold in the United States. Most of this LPG, that is, propane for heat and other home purposes, is used during the winter. The petrochemical industry uses about 38 percent of the LPGs in the manufacture of plastics. Farms and factories use another seven percent each. Farms use the most LPGs in the fall, but factories use a steady amount year-round. Transportation accounts for only three percent.

Benefits of LPG

LPG is a good fuel for internal combustion engines. LPG is no more dangerous than gasoline when contained in a fuel tank. Because LPG becomes liquid easily, it is possible to put it in pressurized tanks for storage and transport. People can keep tanks of LPG in their yards, and tanker trucks can deliver it to rural areas that are not served by natural gas companies.

Propane is an excellent fuel for automobiles and is becoming one of the most popular alternative fuels. Propane vehicles produce between 30 and 90 percent less carbon monoxide and 50 percent fewer smog-producing pollutants than gasoline-powered vehicles. In the early 2000s there were about 350,000 propane-powered vehicles in the United States and about four million in the world. These vehicles include cars, vans, pickup trucks, buses, and delivery trucks. The U.S. Department of Energy has encouraged consumers to consider using propane-powered vehicles.

In many ways propane is superior to electricity and to other fuels. It does not produce nearly as much pollution as gasoline or coal. Propane furnaces are more efficient at heating and release fewer air pollutants than heaters powered by electricity or fuel oil. Propane fireplaces are cheaper and less polluting than wood-burning fireplaces, and they can be turned on and off with a switch. Many professional cooks prefer propane stoves to electric stoves because they produce heat instantly and are easier to control. Moreover, propane appliances will still work during power outages, unlike electric appliances.

Drawbacks of LPG

LPG is more expensive to produce than gasoline. It is not widely available, so it can be difficult to refuel a car that runs on LPG, although in the early 2000s this situation was improving. It can be difficult to find an LPG-powered vehicle because not many are made. Propane-powered vehicles usually have a slightly lower driving range than gasoline-powered vehicles because the energy content of propane is lower than that of gasoline.

LPG is highly explosive. It is important to maintain propane appliances in good condition and have them inspected regularly. Consumers should find out where gas lines run under their yards so they can avoid striking them with shovels or other hard metal objects. Anyone who smells a propane leak should immediately evacuate the building and call the fire department. No one should flip light switches, turn on other electrical appliances, or use the telephone if near a propane leak.

Impact of LPG

LPG emissions of nitrogen oxides, carbon monoxide, hydrocarbons, and particulate matter are very low. LPG releases almost no emissions through evaporation, as gasoline and diesel fuel do. Engines that run on LPG are quieter than those that run on gasoline. LPGs do not cause carbon to accumulate inside machinery.

Economically, because propane and LPGs are produced as a by-product of natural gas and petroleum refining, their prices are directly tied to petroleum and natural gas prices. Prices for LPGs fluctuate (go up and down) according to seasonal demand. They are usually most expensive in winter, when people are using them for heat. Prices also vary by distance from source, so that consumers who live far away from sources of LPGs often pay more for them than consumers who live close by. Automobile manufacturers do not build LPG-burning cars because LPG is more expensive than gasoline.

On a societal level, LPG is invaluable to people in rural areas because it is a source of power that can be transported to areas not otherwise served by natural gas and electricity.

Issues, challenges, and obstacles in the use of LPG

Since LPG comes from the production of petroleum and natural gas, when those supplies run out, so will LPG. At the beginning of the twenty-first century many organizations are trying to encourage consumers to use more propane and LPGs as fuel for their homes or vehicles, and interest in LPGs has increased somewhat as people become concerned about the environment. In order for more people to use LPGs as fuel for transportation, companies will have to make it easier to refuel the vehicles and less expensive to buy them.

METHANOL

Methanol is a kind of alcohol that can be used as fuel. It is also called methyl alcohol and is used primarily in industry and in racecars. Some people hope it can be used to power fuel cells.

Methanol is a clear, colorless liquid with a distinctive odor. Methanol used to be called wood alcohol because people made it by burning wood and condensing the vapors that emerged. The ancient Egyptians created methanol in this way and used it to embalm mummies. Robert Boyle (1627–1691) isolated methanol in the 1660s, and Pierre Eugène Marcelin Berthelot synthesized it in about 1860. In the twenty-first century methanol usually is produced from natural gas. It may be possible to use coal or wood to produce methanol in order to avoid using natural gas resources.

Current and potential uses of methanol

Methanol has several uses. Chemists use it to manufacture plastics and formaldehyde, which is used to preserve organic matter. It is useful as a solvent and as antifreeze. Methanol also can be used to power fuel cells, such as those in cellular telephones or laptop computers, and to manufacture the fuel additive MTBE (methyl tertiary-butyl ether).

Do Not Drink the Methanol

The alcohol people drink in beer, wine, and whiskey is ethyl alcohol, or ethanol. Methanol, although it is a type of alcohol, is not the sort of thing anyone would want to drink. Drinking even a small amount can cause blindness. Drinking a larger amount can kill a person.

Automakers have experimented with using methanol as a fuel for cars, either alone or mixed with gasoline. A mix of 85 percent methanol and 15 percent unleaded regular gasoline (called M85) emits only half the pollutants of gasoline alone. Between 1978 and 1996 several automobile manufacturers made demonstration vehicles that could use both M85 and regular gasoline. Two companies offered these fuel-flexible vehicles for sale to consumers in 1995 and 1996. Methanol is a popular fuel for race cars largely because methanol fires can be put out with water, which makes it safer than gasoline.

Benefits and drawbacks of methanol

When used as an automobile fuel, methanol produces fewer emissions and has better performance than gasoline. It is also less flammable. Methanol can be made from a variety of substances, including natural gas, coal, and wood. Use of methanol could reduce dependence on petroleum. Methanol can easily be made into hydrogen so it has potential as a fuel source for hydrogen fuel cells.

However, methanol has several drawbacks as a fuel. The flame produced by burning methanol is colorless and almost invisible, which makes it dangerous for people working near it. Methanol vapors are poisonous and can burn skin. People who handle methanol without adequate protection can absorb it through their skin or lungs and quickly become ill, because methanol is highly poisonous.

Methanol is also more expensive to produce than gasoline, which makes methanol-gasoline mixes more expensive than plain gasoline. Anyone who owns a methanol-powered vehicle has a hard time finding a place to refuel. Automobile manufacturers stopped making methanol-powered vehicles in 1998, switching their attention to ethanol instead.

Impact of methanol

Methanol produces fewer greenhouse gases than gasoline. Vehicles powered with mixed gasoline and methanol emit just one-half the smog-forming pollutants that a comparable gasoline-powered vehicle emits. The formaldehyde it produces when it burns, however, is quite poisonous.

Many industries use methanol in their daily business. Because most methanol is made from natural gas, changes in natural gas prices affect methanol prices. Some factories that produce methanol stop production if natural gas prices go too high, a practice that can cause methanol shortages.

Issues, challenges, and obstacles in the use of methanol

Many people believe methanol has potential as a fuel. Federal and state governments have passed laws encouraging the development of alternative fuels such as methanol. The California Energy Commission has encouraged car manufacturers to experiment with methanol since 1978. Twenty-five years of experimenting did little to increase public support for using methanol as a fuel. As of 2005 most car manufacturers had abandoned methanol research.

Japanese cellular telephone manufacturers have been developing fuel cells powered by methanol. They hope that by 2007 people will be able to provide hours of power for their cellular telephones by squirting drops of methanol into them. The main drawback to this technology is the need to carry flammable methanol in public places, such as on airplanes. Researchers hope that this technology will have a wider application in the near future.

METHYL TERTIARY-BUTYL ETHER

Methyl tertiary-butyl ether, or MTBE, is a substance added to gasoline to make it burn more completely and produce fewer polluting emissions. It has been added to gasoline in the United States since the late 1970s. In the 1990s communities discovered that MTBE was getting into their water supplies, which led to a movement to eliminate MTBE use.

MTBE is a chemical compound made of methanol and isobutylene. At room temperature, MTBE is a colorless liquid that dissolves easily in water. It is volatile (or unstable) and flammable. It has a strong odor, and small amounts of it can make water taste bad. MTBE is an oxygenate, which is a substance that raises the oxygen content of another substance. MTBE is used to raise the oxygen content of gasoline.

Current and potential uses of MTBE

MTBE, used as a fuel additive, increases the octane level of gasoline and reduces emissions of carbon monoxide and pollutants that form ozone. The U.S. Clean Air Act was passed in 1963 and updated in 1970 and 1990, requiring people in certain areas to use oxygenated gasoline. MTBE is one of the least expensive oxygenates, so most oil companies chose it as a fuel additive. Gasoline with oxygenates added to it is sometimes called reformulated gasoline, or RFG. At the end of the twentieth century about 30 percent of the gasoline sold in the United States was RFG, and MTBE was the oxygenate most commonly mixed into it. MTBE is the primary oxygenate because it is relatively inexpensive.

Benefits and drawbacks of MTBE

MTBE blends easily with gasoline, and it can be shipped through existing pipelines. Gasoline with MTBE mixed into it burns more cleanly than plain gasoline, reducing tailpipe emissions. This has resulted in an improvement in air quality. The U.S. Environmental Protection Agency estimated that the addition of MTBE to gasoline reduces toxic chemical emissions by twenty-four million tons a year and smog-forming pollutants by 105 million tons.

MTBE dissolves easily in water, which can pose a hazard. When gasoline tanks or pipelines leak above or below the ground, the MTBE can dissolve in groundwater and travel to water supplies. Urban runoff, rain, motorboats and jet skis, and car accidents can all result in gasoline and MTBE getting into groundwater. Gasoline tends to stick to soil so it does not travel very far when it is spilled, but MTBE moves freely with water and can easily contaminate water supplies. It does not break down in the environment, so it can stay in groundwater for years.

Some people fear that MTBE causes health problems. Research animals exposed to large amounts of MTBE have developed cancer and other health problems. So far researchers do not believe that MTBE in gasoline poses any major health risks to humans. Researchers do, however, believe MTBE may cause cancer in people who drink water contaminated with large amounts of it.

Impact of MTBE

The use of MTBE in gasoline has improved air quality in the United States since 1995. But MTBE has gotten into the groundwater in some areas. This happens easily when gasoline leaks out of storage containers or is spilled during transport. Rain can carry MTBE into shallow groundwater, and it can then get into deeper water supplies. MTBE can make water undrinkable. Some states have set limits on the amount of MTBE allowed in drinking water. Most public water systems must monitor their water supplies for the presence of MTBE.

Although MTBE can spread through the ground and water very easily, it does not break down easily. Getting MTBE out of water is difficult, so once it has polluted a water source, MTBE can be very hard to clean up. In 1996 the city of Santa Monica, California, found that two wells supplying the city's water were contaminated with MTBE and that levels of MTBE were increasing. After discovering more areas contaminated with MTBE, the state issued an order requiring that MTBE be removed from all California gasoline by the end of 2003.

On an economic level, MTBE is one of the least expensive and most convenient fuel additives. A huge amount of MTBE is produced in the United States. In 1999 more than two hundred thousand barrels were produced every day. Production of MTBE is very profitable, but cleaning MTBE out of the U.S. water supply is very expensive. MTBE has caused a number of lawsuits over cleanups that have cost both cities and oil companies huge amounts of money.

Issues, challenges, and obstacles in the use of MTBE

Many U.S. states have decided that the risks associated with MTBE are too great. Following California's lead, many states have called for MTBE to be phased out completely by 2014. A proposed $2 billion may be spent between 2005 and 2013 to help MTBE manufacturers switch their operations to some other substance.

CONCLUSION

In the early 2000s most of the world is utterly dependent on fossil fuels for its energy needs. A number of nations are deeply concerned about this dependence because the use of fossil fuels contributes to air pollution and sometimes leads to strife between nations, and because the supply of some types of fossil fuel is likely to run out in the not-too-distant future. Many governments have begun looking for ways to end their dependence on oil, by exploring alternative sources of energy and developing systems of public transportation.

■ ■ ■

For More Information

Books

Freese, Barbara. *Coal: A Human History.* New York: Perseus, 2003.

Gelbspan, Ross. *Boiling Point: How Politicians, Big Oil and Coal, Journalists and Activists Are Fueling the Climate Crisis.* New York: Basic Books, 2004.

Leffler, William L. *Petroleum Refining in Nontechnical Language.* Tulsa, OK: Pennwell Books, 2000.

Web Sites

''Alternative Fuels.'' U.S. Department of Energy Alternative Fuels Data Center. http://www.eere.energy.gov/afdc/altfuel/altfuels.html (accessed on July 20, 2005).

''Black Lung.'' United Mine Workers of America. http://www.umwa.org/blacklung/blacklung.shtml (accessed on July 20, 2005).

''Classroom Energy!'' American Petroleum Institute. http://www.classroom-energy.org (accessed on July 20, 2005).

''Oil Spill Facts: Questions and Answers.'' *Exxon Valdez* Oil Spill Trustee Council. http://www.evostc.state.ak.us/facts/qanda.html (accessed on July 20, 2005).

''The Plain English Guide to the Clean Air Act.'' U.S. Environmental Protection Agency. http://www.epa.gov/air/oaqps/peg_caa/pegcaain.html (accessed on July 20, 2005).

Bioenergy

INTRODUCTION: WHAT IS BIOENERGY?

Bioenergy is renewable energy produced by living things like plant matter or by the waste that living creatures produce, such as manure. These living things and their waste products are called biomass. Biomass is organic matter (which comes from living things), just like fossil fuels (coal, oil, or natural gas, which are formed in the earth from plant and/or animal remains), but it is much more recently created and is renewable on a time scale that is useful to humans. Fossil fuels take millions of years to form. During this time they accumulate large amounts of carbon, which is returned to the atmosphere during burning. Plants grow continuously, animals constantly produce manure, and people throw away waste material all the time. Using these items for fuel does not deplete them because they are always being made.

For this reason, many experts believe that bioenergy will be a major source of power in the future. Besides being renewable, many kinds of bioenergy are considered less polluting than fossil fuels. They can be used as direct substitutes for fossil fuels, powering diesel or gasoline engines, heating buildings, and producing electricity. They can be made and used locally, which can make individual areas more self-sufficient and less reliant on foreign suppliers for energy. Bioenergy is created by using biofuels. Biofuels are made from sources of biomass including wood, plant matter, and other waste products. These sources can then be turned into biofuels. There are three types of biofuels: solid, liquid, and gas.

Words to Know

Anaerobic Without air; in the absence of air or oxygen.

Biodiesel Diesel fuel made from vegetable oil.

Bioenergy Energy produced through the combustion of organic materials that are constantly being created, such as plants.

Biofuel A fuel made from organic materials that are constantly being created.

Biomass Organic materials that are constantly being created, such as plants.

Distiller's grain Grain left over from the process of distilling ethanol, which can be used as inexpensive high-protein animal feed.

Feedstock A substance used as a raw material in the creation of another substance.

Flexible fuel vehicle (FFV) A vehicle that can run on a variety of fuel types without modification of the engine.

Infrastructure The framework that is necessary to the functioning of a structure; for example, roads and power lines form part of the infrastructure of a city.

Types of bioenergy

Biofuels come in all three forms of matter: solid, liquid, and gas. Solid biofuels are solid pieces of organic matter that release their energy through burning. Solid biofuels include the following:

- Animal waste (dung or manure)
- Bagasse (plant waste left after a product like juice or sugar has been removed)
- Charcoal
- Garbage
- Straw, dried plants, and the shells of grains
- Wood

Liquid biofuel is any kind of liquid derived from matter that has recently been alive and that can be used as fuel. Types include the following:

- Biodiesel, which is diesel fuel made of vegetable oils and animal fats instead of petroleum.
- Vegetable oil fuel, including straight vegetable oil, or SVO, and waste vegetable oil, or WVO.
- Ethanol and other alcohol fuels, which are made from corn, grain, and other plant matter and can be mixed with or substituted for gasoline.
- New fuels, such as P-Series fuels, which combine ethanol, natural gas traces or leftovers, and a substance made from garbage.

Whale Oil

Whale oil was an important liquid biofuel in the eighteenth and nineteenth centuries. Whalers traveled the world's oceans searching for right whales and sperm whales in order to kill them and remove the oil from their bodies. This oil was used to light lamps and to make candles, cosmetics, and drugs. People still hunted whales in the twentieth century, and new uses were developed for whale oil. However, synthetics and fossil fuels replaced whale oil for almost all purposes by the mid-twentieth century. They were cheaper and more plentiful, especially as whales were hunted to near extinction and became more difficult to find and catch. Most of the world's countries in the twenty-first century have declared whaling, and the taking of whale oil, illegal.

Biofuel can also come in the form of a gas, or biogas, particularly that is emitted (given off) by decaying plants, animals, and manure. This gas is largely methane, which is the main component of the fuel natural gas. Most methane used in 2005 comes from fossil fuels, but scientists are currently researching ways to collect methane from decaying garbage. Scientists are also investigating the possibility of using biofuels to generate hydrogen, which could then be used in fuel cells. Gasification of solid biofuels (transforming their energy into natural gas) is also a possibility.

Historical overview: Notable discoveries and the people who made them

People began experimenting with bioenergy in motors in the mid-1800s. In 1853 scientists used a chemical process with vegetable oil that created biodiesel. Rudolf Diesel (1858–1913), inventor of the diesel engine, gave a speech in 1912 in which he suggested that vegetable oil fuels were destined to become as important as petroleum and coal. However, diesel engine manufacturers in the 1920s geared their engines to run on thicker petrodiesel (diesel made from fossil fuels) because it was cheaper than biodiesel at the time. As a result, manufacturers ignored vegetable oil fuels for most of the twentieth century. Nevertheless, a few people used biodiesel and vegetable oil fuels throughout the 1900s.

A view of ethanol storage tanks that are being constructed on the northern side of the port of Santos, Brazil, September 3, 2004. Brazil is the world's largest producer and exporter of sugar and ethanol. © *Paulo Whitaker/Reuters/Corbis.*

Ethanol, too, generated interest in the early 1900s. Henry Ford (1863–1947) believed ethanol made from grain would be a valuable fuel. No one used much ethanol, however, until the oil embargo of 1973 led to an oil crisis. Convinced that the world was running out of oil, some people decided to use ethanol instead of gasoline. The movement was small in the United States and focused mainly in corn-growing states, but it became a big business in Brazil, which had ample sugarcane to use in making ethanol. Ethanol-burning automobiles were popular in Brazil until the late 1980s, when oil prices came down and sugar prices went up.

In the late 1900s, as people grew increasingly concerned about the limited supply of fossil fuels and the pollution caused by burning them, scientists and consumers once again turned their attention to bioenergy. In the 1990s France began producing

The 1973 Oil Crisis

On October 17, 1973, the nations that belonged to the Organization of Petroleum Exporting Countries, better known as OPEC (OH-pek), announced that they would no longer sell petroleum to nations that had supported Israel in its fight with Egypt. These nations included most of Western Europe and the United States. Oil suddenly cost four times more than it had the month before. Gasoline appeared to be in short supply, and nations began limiting people's access to fuel. The government of the United States realized how dependent it was on Middle Eastern oil and responded by increasing efforts at U.S. oil exploration and extraction. The crisis spurred a new interest in fuel economy and alternate sources of energy. The national speed limit was reduced to 55 miles per hour, Daylight Savings Time was lengthened to save electricity, and the Department of Energy was created.

biodiesel fuel locally, using rapeseed oil to make the fuel. By the end of the twentieth century a large number of French vehicle manufacturers were producing vehicles intended to use some biodiesel in their fuel mixes. Increasing numbers of ethanol fuel plants were being built in the early 2000s. Biofuels may become big business in the twenty-first century as the supply of fossil fuels dwindles and the price of fossil fuels goes up. Biofuels such as biodiesel are increasingly part of European union legislation, so there is much pressure to develop and use them.

How bioenergy works

Biofuels work by burning either directly (such as putting wood logs on a fire) or indirectly as through an engine. They are similar to fossil fuels, which also release their energy when they burn. Biofuels are the alternative fuels most similar to fossil fuels. In many cases they function as direct replacements of, or supplements to, fossil fuels.

The internal combustion engine

Gasoline is the main fuel used in automobiles, which are powered by internal combustion engines. The basic principles of internal combustion have not changed in over one hundred years. They are the same whether the fuel is petroleum-based or biofuel. An internal combustion engine burns a fuel to power pistons, which make the engine turn. An engine contains several cylinders (most cars have between four and eight) that make the whole engine move.

One complete cycle of a four-stroke engine will turn the crankshaft twice. The crankshaft is a shaft connected to a crank that turns and moves the pistons in an engine up and down. A car engine's cylinders can fire hundreds of times in a minute, turning the crankshaft, which transmits its energy into turning the car's wheels. The more air and fuel that can get into a cylinder, the more powerful the engine will be. An engine using methanol is a bit different than one using petroleum or propane, but the concept is similar.

Stoves, campfires, and grills

The simplest technology using solid biofuels is a fire, such as a campfire, which consists of a pile of sticks, logs, or animal dung set on fire. There are many ways to arrange the pile of sticks, logs, or dung for safety and efficiency of burning, but basically the construction of a fire is simple and does not require the addition of complex equipment. Charcoal is often the preferred fuel because some of the other fuels give off smoke that can be harmful to the environment.

There are also devices that make it easier for people to use fires for heat or cooking. Grills placed on top of a fire, or devices that can hold a fire in a bowl with a grill on top of that, make it easy to cook food. Woodstoves come in a variety of styles. Some woodstoves make it possible to heat a large house with a small fire. Others contain both stove tops and ovens for cooking flexibility.

Gas pipes

Gaseous fuels travel through pipes from the place where they are produced to the place where they burn. In London in the 1800s, pipes delivered biogas from the sewers to street lamps. In 2005 some dairy farmers collect biogas from fermenting tanks of manure and run it to their appliances through pipes. The biogas can be lit at the pipe's end, powering a light, a stove, or another appliance.

Current and future technology

Biofuels are already widely used in many parts of the world. Germany, Britain, France and Brazil all use biofuels in different ways. Biodiesel and ethanol are increasingly common. Scientists are working to develop new technologies that can take advantage of currently inaccessible sources of bioenergy.

Vegetable oil was one of the first fuels used in internal combustion engines. Today most vegetable oil is consumed in the form of biodiesel, which functions almost exactly like diesel made from petroleum, called petrodiesel. Vegetable oil, either new or used, can be used for fuel by itself in diesel engines, though the engines must be modified for this to work well. In the twenty-first century large companies are taking more of an interest in biodiesel; commercially prepared biodiesel is becoming more widely available, either straight or mixed with petrodiesel.

Ethanol, which is the same alcohol used in alcoholic beverages, has a long history of use as a fuel. Other alcohols, such as methanol, can also be used as fuel. Ethanol is easy to make from corn, grain, sugarcane, or other plant material. Ethanol can be mixed with gasoline to power internal combustion engines. Normal cars can use small amounts of ethanol in their fuel. Flexible fuel vehicles (FFVs) can use fuel that is nearly all ethanol. (Few vehicles in the early twenty-first century can use straight ethanol with no gasoline in it.) In some parts of the world, ethanol is routinely mixed in fuel, reducing the use of fossil fuels.

Scientists are also working to develop new types of fuels. P-Series fuel is a fuel that is made from a combination of ethanol, the leftovers from natural gas processing, and a substance made from garbage. It works in flexible fuel vehicles and appears to be a stable substitute for gasoline. Whether these fules are pollutants or not has not been concluded. Some scientists believe that they are non-polluting, but others believe that they give off significant nitrogen oxide emissions.

Benefits and drawbacks

Biofuels appear likely to furnish at least some of the world's energy needs. There are many good reasons to use biofuels:

- They are environmentally much cleaner than fossil fuels, producing less air pollution and consuming materials that would otherwise be considered garbage.
- They are renewable; the supply of biofuels is less likely to run out, while the supply of fossil fuels probably will.
- They can be made locally using local materials.
- They can be flexible, easily mixed with other fuels.
- They can be cheaper than fossil fuels and will certainly become less expensive as the price of fossil fuel rises.

Singer Willie Nelson poses with a pump for Biodiesel fuel. Nelson, along with his business partners, are marketing a brand of clean-burning biodiesel fuel. The fuel is made from vegetable oils, mainly soybeans, or from animal fats, that can be used without modification to diesel engines. *AP Images.*

- Ethanol and biodiesel are better for car engines than fossil fuels. They can be used as additives to improve performance even if they are not the main fuel source.

But biofuels are not without some disadvantages:

- To make large amounts of biofuels would require cultivating more land than is currently farmed. This could be a very large problem to try to overcome.
- Some kinds of biofuels require modifications to vehicle engines.

Flexible Fuel Vehicles

Flexible fuel vehicles, or FFVs, are vehicles that can run on various kinds of fuels, such as ethanol-gasoline blends, methanol, gasoline, P-Series fuels, or combinations of those, without having to be physically modified. The engine contains sensors that identify the type of fuel and adjust the timing of the spark plugs and fuel injectors to provide the optimum combustion.

In 2005 some common FFVs included the Ford Explorer, the GM Yukon, and the Mercedes-Benz C320. The owner's manual of the car states if the vehicle is indeed an FFV. Most FFVs are in the sport utility vehicle or light truck category. Sedans that are FFVs are usually made specifically to be fleet vehicles, one of many identical cars owned and used by a large company or organization.

- Making biodiesel at home or processing vegetable oil for home use is messy and inconvenient.
- Biofuels are not widely available.
- Some biofuels still require the use of fossil fuels; for example, most vehicles must have some gasoline mixed into ethanol to work and cannot run on ethanol alone.

Environmental impacts

Bioenergy is less polluting than fossil fuel–produced energy in respect to carbon dioxide. Biofuels contain carbon that only recently was in Earth's atmosphere, so the carbon dioxide released through burning them does not add to total carbon dioxide in the air. Fossil fuels, however, contain carbon that was removed from the atmosphere millions of years ago, and they emit large amounts of extra carbon dioxide when they burn. Replacing some fossil fuels with biofuels may help ease global warming, lessen air pollution, and clean the world's air.

Bioenergy, however, may be a contributor of formaldehyde to urban air. Biodiesel fuels are potentially high emitters of nitrogen oxides, which are a major component of smog. People with respiratory illnesses and small children are most affected by these air pollutants.

Biofuels are renewable. They come from plants and other currently growing organic material, so it is possible to generate new ones constantly. This makes them more environmentally

The power from this power plant is generated from the methane of the manure of the cattle grazing in the foreground. ©*Charles O'Rear/Corbis.*

appealing than fossil fuels, which are, for all practical purposes, not renewable and are even in the process of being depleted.

Biofuels can use waste for feedstocks (starting materials). For example, waste vegetable oil from fast food restaurants or potato chip factories can be turned into biodiesel. This prevents the waste material from being disposed of in a landfill.

On the other hand, biofuels require large amounts of land to be cultivated and harvested. This can cause major environmental problems, such as habitat destruction and fertilizer runoff. Farmers use large amounts of fossil fuels to grow crops such as corn, which decreases the value of the energy made from those crops. In some cases, producing biofuels such as ethanol actually uses more energy than the ethanol yields.

Economic impacts

In the early 2000s production of biofuels increased very rapidly. In the United States production of ethanol rose 30 percent each year between 2000 and 2005. During the same period Germany's production of biodiesel increased by 40 to 50 percent annually. France planned to triple its output of ethanol and biodiesel between 2005 and 2007, while Britain built two major biodiesel plants during the first few years of the century. As of 2005 China had built the world's largest ethanol plant and intended to build another just like it. A Canadian company planned to build a plant to make ethanol out of straw.

The reason for this increase was simple. Biofuels had previously been more expensive than fossil fuels, making them uneconomical during most of the twentieth century. Some people had supported biofuels all along because they wanted the world to use fuels that they believed were not as damaging to the environment as fossil fuels, and they persuaded governments to back them. But in the early 2000s it became clear that biofuels also made good economic sense. The price of fossil fuels went up, making biofuels comparatively cheaper. Depending on location, biofuels even became cheaper in real terms, that is, without governmental supports.

For individual consumers, biofuels can be more or less expensive than fossil fuels depending on how they are used. People who make their own biodiesel using free waste vegetable oil from restaurants spend very little money on fuel, though they do spend a certain amount of time in the pursuit of energy. Wood heat can be less expensive than electrical or gas heat. In the past, purchasing biofuels was usually more expensive than purchasing fossil fuel equivalents. That is changing in the twenty-first century, and more people are finding that biofuels make economic sense.

Societal impacts

One of the biggest impacts that biofuels can have on society is increased self-sufficiency for areas and individuals that use them. Individual consumers and most nations do not have fossil fuels

A Cambodian villager stands in front of her gas stove powered by a biological gas digester. It converts human and animal manure into an enviornmentally friendly fuel. *AP Images.*

readily available. People and countries must buy their oil, gas, or coal from large companies that drill and process them. Consumers are vulnerable to changes in the price or supply of oil. Using biofuels allows people or nations to make their own fuel on the spot. This is especially useful to developing nations that have a large need for energy but do not have much money to buy fossil fuels.

Barriers to implementation and acceptance

Few people used biofuels during the twentieth century because fossil fuels were readily available and inexpensive. By the early twenty-first century biofuels were becoming more attractive to large companies, which suddenly saw biodiesel and ethanol as potential sources of profit. Oil and gas companies still have little interest in pursuing sources of bioenergy, and their influence on national and state governments could prevent biofuels from being used in public transportation fleets.

Most people still know little about biofuels and so do not seek them out. Biofuels are not readily available in many places, so it is difficult for people to use them. Few people want to go to the trouble of making their own biodiesel or modifying their car engines to run on vegetable oil. As biofuels become more commercially available and user-friendly, consumers are likely to adopt them in increasing numbers.

SOLID BIOMASS

Solid biomass was the first fuel humans ever used. Prehistoric humans used wood and animal dung to make their first fires, over which they cooked food and kept themselves warm. Ever since, solid biomass has played an important role in human energy needs.

Biomass energy is energy derived from solid organic matter other than fossil fuels. This can include charcoal, wood, straw, hulls of grains, animal manure, and bagasse (solids left over from the processing of sugarcane or fruit). The fuel can be used directly as for fire or used to power other devices such as electrical generators.

Solid biomass fuel can be used as it is found, but it often benefits from some processing to make it drier or denser than it is in nature. For example, the process of making charcoal transforms wood into a dry substance that is nearly pure carbon. Removing impurities can also improve efficiency.

Whether solid biomass is renewable or not depends entirely on how rapidly it is used. Wood met human energy needs for millennia, but once a forest is completely cut down, it becomes useless until the trees grow back, if they do.

Current uses of solid biomass

Solid biomass is still widely used around the world. Most developed nations have moved away from using solid fuels for their day-to-day energy needs, favoring more efficient and readily available fossil fuels, but in much of the world wood for the fire is still a daily necessity. People in developed countries still use solid biomass as a source of fuel for some purposes.

Animal waste

Animal feces, also called manure or dung, is an important source of fuel around the world. Manure contains large amounts of carbon and nutrients that can be used as fertilizer, but it also contains ample plant fiber that will burn. Dried dung is widely used as fuel for fires in areas where there are many animals. People use dung from cattle, buffalo, horses, llamas, kangaroos, and other

The Problem with Poop

Manure is a big problem for anyone who raises large numbers of farm animals such as cattle. A herd of cattle produces a tremendous amount of manure. Modern farmers gather the manure from barns or pastures and collect it in large heaps. Manure, like any pile of organic matter, gets hot as it decomposes. It can get so hot that it catches on fire, which results in a dangerous situation and very stinky smoke. Piles of poop also breed flies, which can spread disease. Though manure has many uses, on large farms it quickly becomes too much of a good thing. That is why scientists are investigating uses for manure such as the production of biogas.

creatures. Biomass sources such as garbage and manure can be allowed to decay to produce methane, or natural gas.

Bagasse

Bagasse is the solid material left behind after removing a product from its source, such as juice from oranges or grapes and the sugar from sugarcane. About 30 percent of the sugarcane is left over after processing, and this solid fibrous material has long been used as fuel. In the earliest days of sugarcane processing, the bagasse was used as fuel for the sugar mill; in some cases, the processors would not extract as much sugar as they could from the bagasse so that they would have more bagasse left over to burn.

Bagasse is now an important source of fuel in Brazil. Brazil expanded its sugar industry in the 1970s to make sugar to produce ethanol. The ethanol plants use bagasse to power their machinery. Brazilian sugar growers sell excess bagasse to other industries, such as juice and vegetable oil factories, which burn it instead of fuel oil. This saves the nation several million dollars a year in oil import costs.

Charcoal

Charcoal is a black combustible material made by removing the water and volatile substances from wood or other organic materials. It consists almost entirely of carbon, usually between 85 and 98 percent. The main reason to make wood into charcoal is to make it burn hotter and more efficiently. Wood contains a great deal of water, which cools the fire; volatile compounds such as

Two bagasse-burning Trankil Sugar Mill steam locomotives, numbers 3 and 4, in operation in Java, Indonesia. The bales on the back of the locomotives are pressed bagasse fuel from sugar cane waste. © *Colin Garratt; Milepost 92/Corbis.*

methane and hydrogen; and tars. To make charcoal, wood is buried to prevent oxygen from reaching it and allowing it to catch fire and then baked at a moderate heat for many hours. The impurities burn off in smoky clouds. Commercial charcoal manufacturers add borax to bind (hold together) the charcoal, nitrate to help it catch fire, and lime to color the ashes white.

Most twenty-first century Americans use charcoal only on outdoor grills, but charcoal has a long history of use. Bronze Age Europeans five thousand years ago used charcoal to melt metals. Blacksmiths used charcoal because it produced more heat than wood, important for heating metal. Charcoal was fuel in glassmaking and cooking. Artists use charcoal to create soft gray or black lines that blend easily. Charcoal is an ingredient in gunpowder. Charcoal in the metallurgy industry (metal industry) has now largely been replaced by fossil fuels such as coke and anthracite coal.

Compost

Compost is organic material that has decomposed and turned into humus (material that results from partially decomposed plant and animal matter). It is added to soil as fertilizer and to improve the soil's structure. Plant matter that falls to the ground, such as leaves from trees, naturally decomposes and becomes part of the soil. Composting is the practice of consolidating this matter and controlling the conditions under which it decomposes, which speeds up the decomposition greatly. A compost bin or pile can contain dried leaves, green plant matter, table scraps such as vegetable skins, animal manure, and even paper. The mix needs water and oxygen to decompose properly. Microbes and insects such as ants break down the organic matter and turn it into a substance that looks very much like dirt. Though a simple trash heap will eventually produce usable compost over many months, a skilled composter will use techniques that make the pile grow very hot, killing seeds and germs and producing usable compost in just a few weeks.

Compost is not itself an energy source, but it can be a valuable replacement for fertilizers made with fossil fuels like natural gas. Farmland enriched with compost is more fertile than uncomposted land because the nutrients from the compost become part of the soil. Organic gardeners use compost to recycle yard and table waste and to make their soil richer. About one-third of landfill space is occupied by yard waste and table scraps. Putting yard waste and table scraps into compost saves landfill space by turning those materials into dirt.

Garbage

Garbage is usually seen as a problem—as waste material that must be dumped somewhere, but preferably not close to anyone's home. Some scientists, however, have been experimenting with ways to turn garbage into fuels or useful substances. Some types of garbage can be converted into biogas, which can be used as fuel. Garbage is also a component of P-Series fuel.

Straw, dried plants, and shells of grains

There is some possibility that dried plant matter could be used to manufacture ethanol. Making ethanol from this kind of cellulose (cellular material in plants) is more difficult than making it from sugarcane or grain. Straw and hulls do not contain as much sugar, and it is more difficult to remove the sugar from them, but it is possible. These dried substances can also be made into compost or converted into biogas. They are usually not the best fuels for fires because they burn quickly and cannot produce long-lasting heat.

Ancient Central Heating: The Hypocaust

The ancient Romans used wood to create central heating for their homes. They used a system called a hypocaust. A hypocaust was a structure of tunnels under the floor of a building leading up into ducts in the walls of rooms. People would light a fire in the hypocaust, and the warm air would flow through the tunnels and air ducts, heating the building. This system was also used in public baths to heat floors, rooms, and water. A hypocaust was not a practical solution for most people because it required several slaves to feed the fires and remove ashes, and it could only be implemented in buildings made of stone or brick.

Wood

Wood is perhaps the oldest solid biomass fuel. For most of human existence people have burned wood for heat and cooking. In many parts of the world wood is still the primary or the only available source of power. In the United States, wood is still a common source of heat in colder climates where it is plentiful. Some homes have wood stoves that burn wood to heat the house. Others have boilers outside the house that pipe heat into the home. Wood had a brief resurgence in popularity after the 1973 oil crisis.

The residues of wood and other forms of biomass can be used as a source of gaseous fuel. For example, wood residue inside a reactor vessel can be heated to make it break down and produce gas. This gaseous fuel can be burned on the spot as fuel for a turbine or other device.

Benefits and drawbacks of solid biomass

Solid biomass is renewable, at least as long as plants keep growing and are not harvested faster than they can replace themselves. Solid biomass is flexible; a stove that can burn wood can probably also burn charcoal, dung, or other solid matter, though the results may be different. It can be used for simple purposes such as heating a home directly or complex ones such as generating electricity.

Yet solid biomass is only renewable as long as it is not consumed faster than it can be replaced. Solid biomass fuels have much lower energy content than fossil fuels, which means that people using them must acquire large quantities of them to do the same jobs that much smaller quantities of fossil fuels can achieve. Coal, for example, burns much hotter than wood and lasts longer. Anyone using solid biomass for home heating and energy must have access and transportation for large quantities of fuel and must be able to store it until it is needed, such as in a woodpile.

Preparing solid biomass fuels can take a great deal of work. Wood is heavy to move and must be cut into small pieces to fit into stoves or fireplaces. Dung must be collected and carried to wherever it will be burned, and bagasse takes up a great deal of space. A fire fed by solid biomass fuels must be fed regularly or it will go out. Fireplaces and stoves fill with ashes that must be removed from time to time. Ventilation can be a problem, because these fuels all produce smoke. In addition, smoke from woodstoves can be very high in carcinogenic substances.

Environmental impact of solid biomass

Burning wood does not contribute to greenhouse gases because burning wood releases no more carbon dioxide than can be consumed by growing trees. Modern heating stoves are designed to emit few greenhouse gases. Burning wood does produce ash, but ash can be used as fertilizer or in soap making. Bagasse likewise produces few greenhouse gases. However, burning any renewable biomass fuel causes smoke that can seriously cloud the air in the immediate area. Sugarcane cutters often burn cane fields before cutting down the sugarcane. The resulting smoke can linger in nearby towns for weeks. Animal dung causes especially bad air pollution; the World Health Organization estimates that 1.5 million people have died of inhaling air polluted by burning dung.

Issues, challenges, and obstacles of solid biomass

Deforestation (the destruction of forests) is a growing problem around the world. Without enough trees to provide wood, solid biomass fuel will not be a practical source of energy. Tree farming has largely solved this problem in the developed world, but in places where solid biomass fuel is still the main fuel source, lack of trees is a serious problem.

Woodstoves experienced a surge of popularity in the 1970s, after the oil crisis of 1973. Since then other sources of fuel have once again grown in popularity. Solid biomass fuels do not contain as much energy per weight as fossil fuels, so they are not the focus of most research into future energy sources. Wood, charcoal, bagasse, and other solids will probably still be used in the future but only for small-scale purposes such as home heating and cooking. Though biomass fuels have the potential to be a valuable source of energy in some places, such as Brazil's bagasse electricity industry, in most areas they do not seem to be practical sources of large-scale power.

BIODIESEL

Biodeisel is diesel fuel made from renewable sources of carbon such as used vegetable oil or animal fats used in cooking. In diesel engines it can be used as a direct substitute for petrodiesel fuel made from petroleum.

Biodiesel is a clear amber liquid. Its consistency is similar to that of petrodiesel. Biodiesel can be used on its own in a diesel engine or mixed with petrodiesel. Some people mix small amounts of biodiesel into gasoline to decrease its air-polluting qualities.

Biodiesel is usually made out of the vegetable oil that is most readily available in a particular area. In France most commercial biodiesel is made from rapeseed oil. Other kinds of oil used to make biodiesel include palm, mustard, Jatropha, and soybean.

In the United States, soybeans make up the biggest source of biodiesel fuel because they are widely grown. Soybeans are not a particularly good source of biodiesel, but soybean growers have been able to expand the market for soybean-based biodiesel. Rapeseed, mustard, and Jatropha all produce two or three times as much oil as soybeans. Palm oil is an excellent source of oil to make biodiesel, and there has been some research into growing algae to use in making the fuel. Scientists are working on developing crops that produce larger amounts of oil for use in making biodiesel.

Biodiesel users sometimes refer to biodiesel or biodiesel blends by the letter B followed by a number indicating the percentage of biodiesel in the mix. For example, B20 is petrodiesel that contains 20 percent biodiesel. B100 is pure biodiesel.

Vegetable oil into diesel fuel

It is possible to run a diesel vehicle on plain vegetable oil from the grocery store. The first diesel engine ran on straight peanut oil. In diesel engines, however, unprocessed vegetable oil is not very good for the engine because it eventually clogs the filters. In order to keep running the vehicle on vegetable oil, the owner must modify the engine; this is generally true even if the owner mixes the vegetable oil with petrodiesel or kerosene. If the vegetable oil is transformed into biodiesel, however, it becomes so similar to petrodiesel that it can be used in an unmodified diesel engine with no ill effects.

Biodiesel can be made from either new or used vegetable oil or from animal fat. Vegetable oil is the most common feedstock. Waste oil is more difficult to process into biodiesel than virgin oil because it must first be filtered to remove impurities. On the other hand, it is cheaper, often free, and is a good way of recycling a product that otherwise would be thrown away.

How biodiesel is made

Making biodiesel involves joining the fatty acids of the vegetable oil or animal fat into long chains of triglycerides in a process called transesterification. This process converts the oil into long chains of mono-alkyl esters and glycerin. To transform the fats into biodiesel, a processor mixes an alcohol with a lye catalyst (something which causes a chemical reaction faster or at a different rate than it normally would) and then combines the mixture with warm oil. The most common alcohol used in this process is methanol, or methyl alcohol, but ethanol will work as well. The fatty acids float to the top of the mix and are siphoned off as biodiesel, while the glycerin stays at the bottom of the mixing vessel. The biodiesel must then be washed to remove any contaminants that could damage an engine.

Many people make their own biodiesel at home. There are many recipes available, easily found on the Internet. Though biodiesel fans claim that whipping up a weekly batch is no problem, the procedure involves a certain amount of trouble, mess, and danger.

Current use of biodiesel

Biodiesel will work in any diesel engine, with no modifications necessary. This means it can be used as a substitute for petrodiesel fuel. It can be mixed into petrodiesel to reduce emissions, improve engine performance, and clean engine parts.

For many years the only people using biodiesel were enthusiastic environmentalists who made their own biodiesel at home, but that has changed. Commercial suppliers have been making biodiesel and selling it to the public for several years. Biodiesel is widely used in Europe and Asia. France is the world's largest producer of biodiesel. All petrodiesel fuel sold in France contains at least 5 percent biodiesel. In Germany over 1,500 filling stations sell biodiesel, which is less expensive than petrodiesel. The European Union, of which

France and Germany are members, passed legislation to require all member states to mix biodiesel into their petrodiesel. Public transportation fleets are often the first vehicles to adopt the use of biodiesel or biodiesel-petrodiesel blends as their standard fuel.

In the early 2000s biodiesel is becoming more common in the United States. Several states have passed laws requiring biodiesel to be mixed into diesel fuel. Over five hundred commercial fleets use biodiesel. Users include the United States Postal Service, the United States Marine Corps, the National Aeronautics and Space Administration (NASA), the United States Department of Agriculture, numerous state departments of transportation, and the San Francisco International Airport.

The use of biodiesel is increasing rapidly worldwide. In 1998, for instance, 380,000 gallons of commercially manufactured biodiesel were sold in the United States. That amount increased to thirty million gallons in 2004. Biodiesel production is fast becoming a viable economic opportunity and is attracting investors and inventors.

Benefits and drawbacks of biodiesel

Biodiesel has many benefits. It is very easy to substitute for petrodiesel. Employees do not need special training to use it and no equipment needs to be modified. Unlike petrodiesel, biodiesel will not catch fire or explode. It is not poisonous to humans. It is completely biodegradable (capable of being broken down into harmless products). It is environmentally much cleaner than petrodiesel.

In addition, bodiesel is an excellent engine cleaner. It will remove dirt and residue left in a tank and fuel system by petrodiesel. Biodiesel can be added to ultra-low-sulfur petrodiesel to improve its lubricity (ability to reduce friction or rubbing). It makes the diesel fuel flow more smoothly and prevents the accumulation of contaminants within the engine and fuel system.

One reason many people make their own biodiesel is that they take pride in being independent of oil companies and being able to create their own fuel. Many of them save a great deal of money as well, but for many the feeling of independence and environmental virtue is the real attraction.

One major problem with biodiesel is that it is not widely available. France, Germany, and other European countries have many filling stations that sell it, but biodiesel is rare in the United States. For this reason, many people make their own, which itself presents problems. Making biodiesel is time-consuming and can be dangerous. Waste oil must be filtered before it can be used. The chemicals used to make biodiesel are poisonous to humans. Anyone making biodiesel must purchase safety equipment, including gloves, aprons, and respirators, and must have access to a secure work area that children and animals cannot enter.

Another drawback is that biodiesel can be more expensive than petrodiesel, depending on its ingredients. Purchasing new vegetable oil can be expensive, and biodiesel users must often purchase other ingredients and equipment to make the fuel. Converting a diesel engine to run on SVO (straight vegetable oil) can cost money.

Also, biodiesel is not as effective as petrodiesel in cold weather. Both kinds of diesel fuel get cloudy and full of small wax crystals that can clog fuel filters, but biodiesel is more sensitive to this problem than petrodiesel. When biodiesel gets cold enough, it turns into a solid and will not flow at all. Biodiesel made from virgin oil stays fluid at lower temperatures than biodiesel made from waste oil. Most biodiesel users find that they have difficulty with their fuel when temperatures fall below freezing. Some people get around this problem by adding 30 percent petrodiesel to their biodiesel. Others add anti-gel agents to winterize the fuel. Some people worry that biodiesel will decay rubber parts within the fuel system. This can happen, but rubber parts have been uncommon since the 1980s and are easily replaced in any case.

Environmental impact of biodiesel

Biodiesel is much better for the environment than petrodiesel. It is completely biodegradable and non-toxic. It poses no threats to human health. It does not emit the pollutants produced by fossil fuels, which makes it very appealing for areas trying to improve air quality. It does not emit the black smoke that petrodiesel does. It is safe to store and transport. Its flash point (the temperature at which it will catch fire) is over 257°F (125°C), as opposed to 136°F (58°C) for petrodiesel, so it is harder to start a fire with biodiesel.

Making biodiesel is a good way to recycle waste oil that would otherwise end up in a landfill. Though there is a large amount of waste vegetable oil (WVO) produced daily, it is nowhere near the amount of diesel fuel used every day. Likewise, waste animal fat is not nearly plentiful enough to meet major energy needs. Some WVO is already converted into other products, such as soap. Nevertheless, a large amount of WVO and animal fats currently end up in landfills and could profitably be converted to biodiesel.

Economic impact of biodiesel

Biodiesel can function as a substitute for petrodiesel, so economic costs depend partly on what a person or company would be spending on petrodiesel. Nations that use a great deal of biodiesel do not have to purchase petrodiesel from foreign suppliers, which can mean a tremendous savings. The cost of biodiesel to individual consumers varies depending on where they are and how they get it. In Europe biodiesel is widely available and in many places is less expensive than petrodiesel. In the United Kingdom, taxes on biodiesel are lower than those on petrodiesel. Biodiesel is still more expensive than petrodiesel in the United States, but use is increasing and the price is dropping as a result. Several states are considering laws that would require all petrodiesel to include a portion of biodiesel. There are tax credits available to businesses that use biodiesel.

The first users of biodiesel made their own, and this practice is still popular. People who make their own biodiesel using freely donated waste vegetable oil claim to be able to run their vehicles on just a few dollars a month. Integrating biodiesel into an existing petrodiesel infrastructure is not expensive because very few things need to be changed. The equipment is the same, and no training is

Students heading to their school bus, which runs on biodiesel fuel, Wednesday, Sept. 7, 2005, in Morgantown, West Virginia. *AP Images.*

necessary. Because biodiesel is better for engines than petrodiesel, using biodiesel can make an engine last longer and break down less often. Making biodiesel commercially is becoming more profitable as more people purchase it. Thus, producers are taking a much greater interest in making biodiesel for sale as it becomes more profitable.

Issues, challenges, and obstacles of biodiesel

Biodiesel is growing in popularity. In 2006 it is far past the experimental stage and is in the process of being accepted as a mainstream fuel. But public officials and consumers are sometimes resistant to change for various reasons. Public transportation fleets must often coordinate their fuels so that they all use the same ones, which can make it difficult to introduce new fuels. Politicians make promises to various industries, which can also hamper efforts to introduce biofuels.

VEGETABLE OIL FUELS

It is possible to power a diesel engine on plain vegetable oil. This usually requires the engine's owner to modify it slightly. There are two main types of vegetable oil fuels. Straight vegetable oil, or SVO, is exactly what it seems: vegetable oil, just like the kind available in the grocery store. In fact, many people buy vegetable oil from the grocery store to use as fuel. SVO will work in a diesel engine, though for best results the engine needs to be modified. The second type is waste vegetable oil, or WVO, which is oil that has already been used for cooking and can no longer be used for that purpose. Fast food establishments and potato chip factories produce huge amounts of WVO. This oil can be collected, purified, and used as SVO fuel. Waste vegetable oil can also be used as animal feed.

Both SVO and WVO can be used just as they are in engines modified to use them. They can also be mixed with diesel fuel or kerosene to combine the benefits of biofuels with the advantages of fossil fuel. Or they can also be converted to biodiesel.

Current use of vegetable oil fuels

Vegetable oils are mainly used in diesel engines. If the vegetable oil is not converted into biodiesel, which can be used in an ordinary diesel engine, the engine must be modified to get the best results.

SVO can run an engine on its own. So can WVO, which functions just like SVO once it has been cleaned. There are two main

ways to convert an engine to run on SVO. One way is to use a single tank fitted with different filters, temperature controls, injectors, injector pumps, glow plugs, and a fuel pre-heater. Some single-tank systems can run on SVO, biodiesel, or regular petrodiesel. Other vehicles use a two-tank system; one tank holds petrodiesel or biodiesel, and the other contains SVO. The vehicle uses the tank holding diesel to start and warm the SVO and then switches to the SVO to provide power. Using SVO without modifying the engine will gradually result in clogged injectors.

It is possible to use SVO in a diesel engine without modifying it. This is not a practical long-term practice, however. The filters and fuel injectors gradually get clogged up and can cause engine failure.

Some people mix vegetable oil into diesel fuel or kerosene. These blends can contain various proportions of vegetable oil to petrodiesel, mixed according to personal preference and what is available. Though mixed fuel can work in an ordinary diesel

Gasohol 95 is a gasoline extender made from a mixture of gasoline and ethanol. Thailand, whose daily consumption of ethanol is about 66,045 gallons (250,000 liters), plans to raise its daily consumption 12-fold by 2006, which would reach 10 percent of its daily demand for gasoline.
© Chaiwat Subprasom/ Reuters/Corbis.

engine, the best results come from using a two-tank system such as the one that can be used with SVO.

People who use biofuels often see mixes as a poor compromise. The engine still must be modified as if it were running on SVO, and the user is still consuming fossil fuels and emitting pollutants. On the other hand, mixing SVO with petrodiesel or kerosene offers some advantages over straight SVO. It avoids some pollution caused by burning straight fossil fuel, and the engine starts better in cold weather than when it's powered by either biodiesel or SVO.

Benefits and drawbacks of vegetable oil fuels

Using vegetable oil for fuel has many benefits. It is environmentally clean. If WVO is used, it prevents that oil from ending up in a landfill. It is not a fossil fuel, so its use can make regions more self-sufficient and less dependent on foreign sources of oil. People who use SVO as fuel tend to be independent experimenters; they especially enjoy the sense of freedom they get from using fuel that they can acquire themselves.

But using SVO or WVO in engines requires modifying them, which is inconvenient and expensive. SVO is not a direct substitute for diesel, unlike biodiesel, and cannot be used alternately with petrodiesel. Even though using SVO does not require the user to make biodiesel, it still must be prepared before it is burned; WVO especially must be cleaned of all food particles.

Liquid biofuels have a higher viscosity (a level of stickiness) than diesel fuel. This means they do not flow as well in the engine, especially at cold temperatures. Below about 40° Fahrenheit (4.5° Celsius), vegetable oil can solidify, making it useless.

One side effect of using cooking oil in diesel engines is that the exhaust fumes smell like cooking food. Most people do not consider this a major drawback, especially because diesel exhaust fumes also have an odor.

Environmental impact of vegetable oil fuels

SVO is a very clean fuel. SVO mixed with diesel or kerosene is not as clean and still releases the emissions of fossil fuels. Yet it does reduce somewhat the amount of fossil fuels consumed and burned.

In the year 2000 the United States produced over 11 billion liters of waste vegetable oil, most of it from deep fryers in potato chip factories and fast food restaurants. This oil is usually thrown away. Using WVO for fuel is an excellent way of getting rid of

waste oil and avoiding the consumption of fossil fuels. On the other hand, vegetable oil must come from plants, and these plants must be grown. Substituting vegetable oil for fossil fuels will require as much land as possible to be devoted to growing crops that can produce it.

Economic impact of vegetable oil fuels

The economics of using vegetable oil for fuel depend somewhat on whether the oil fuel is new or used. Purchasing new vegetable oil can potentially cost more than purchasing diesel fuel. However, in many cases WVO is free for the taking. Factories and restaurants must pay to dispose of their WVO in the garbage. Therefore, they are often willing to donate it to anyone who wants to collect it. Some enterprising individuals retrieve WVO from local shops and use it in their vehicles, either straight or converted to biodiesel. These people can run their cars for as little as $8 a month, much less than the cost of fueling a gasoline- or diesel-powered vehicle. Even purchasing WVO is inexpensive; in 2003 it sold for about 40 cents a gallon.

Do Not Steal that Oil!

Even though WVO is often freely donated to people who ask for it, it is a bad idea to take WVO directly from a dumpster without asking first. The WVO usually belongs to the company that owns the dumpster, and anyone who takes oil out of it without permission can be charged with stealing. The best approach is to ask individual restaurant owners if they would mind pouring their used oil back into the containers it came in and putting it out for collection by people who want it for fuel. Oil fuel hobbyists claim that Asian restaurants are often a good source of oil because they have the best quality WVO. Hamburger restaurants often have the worst quality WVO. Biodiesel hobbyists also emphasize the importance of maintaining a good relationship with the restaurants that supply them with WVO.

Issues, challenges, and obstacles of vegetable oil fuels

SVO, WVO, and animal fats are popular substances for experimentation. There are many people who would love to be able to run their vehicles and equipment on unmodified cooking oils. As fossil fuels grow more expensive, more commercial enterprises have taken an interest in alternative fuels. Most of this interest, however, seems to be focused on biodiesel, not on SVO. Biodiesel is a much more practical alternative to petrodiesel than SVO because it does not force people to change their vehicles. For that reason, SVO as a fuel by itself is a less likely alternative fuel than biodiesel.

BIOGAS

Biogas is a mixture of gases produced by the fermentation of waste material in anaerobic (without air) conditions. Biogas technology is also called "anaerobic digestion technology." The gases include methane, carbon dioxide, and trace gases such as ammonia, nitrogen, hydrogen, sulfur dioxide, and hydrogen sulfide. Generally the methane content is between 60 and 70 percent. Methane works like natural gas drilled from the ground as a fossil fuel, but unlike natural gas, biogas is renewable. Many people think biogas is an ideal form of energy because it turns waste material into a source of power that produces few pollutants.

Biogas develops in nature all the time. The distinctive smell of swamps is caused by marsh gas, or methane and other gases that develop when vegetation that settles to the bottoms of wetlands is anaerobically digested by bacteria. The manure of cattle in particular contains a great deal of biogas produced by bacteria living in their intestines. These bacteria digest the cellulose in the plant matter that the cattle eat and release methane and carbon dioxide. To collect biogas from manure, a processor collects the manure in a closed tank called a digester. The bacteria digest the cellulose through anaerobic digestion and release methane and other gases into the tank. The biogas can then be collected or piped to wherever it is needed. Biogas can also be made from garbage in landfills or from sewage. Scientists have been developing many different techniques of capturing and using biogas.

Current uses of biogas

Because biogas contains so much methane, it can be used to power appliances that run on natural gas. In many parts of the world biogas is used as a substitute for natural gas, either to run appliances and vehicles or as a source of electricity. A digester on a large dairy farm can produce between four and six million cubic feet of biogas annually, resulting in 124,000 to 198,000 kilowatt-hours of electricity.

Biogas is commonly used in rural areas where there is a ready supply of manure or garbage. In the Netherlands and Denmark biogas is a common source of power. In the United States some dairy farms have begun using biogas systems as a way of managing their increasing manure supplies. In Canada, landfill gas is a major source of energy for electricity generation.

Benefits and drawbacks of biogas

Biogas offers many benefits. It is a good way to get rid of waste materials. The energy it produces is powerful and clean. It does not pollute groundwater or air. Methane can power appliances and vehicles and can be used to generate electricity. Biogas is also quite safe. Homemade biogas does not present any risk of explosion because the gas accumulates slowly and dissipates (goes away) quickly if it leaks instead of pooling on the ground as gasoline does.

But biogas has only one-half the heating value of natural gas. There is not much biogas infrastructure available, so the use of biogas is limited. In addition, using biogas requires the installation of expensive new equipment.

A researcher checks a sealed pot of methane in an electrical generator. The system breaks down organic wastes to create gases which then power an eco-friendly fuel cell generator. One ton of organic waste is enough to generate 580 kilowatts of power, or the equivalent of an average household's electricity consumption over two months. © *Reuters/ Corbis.*

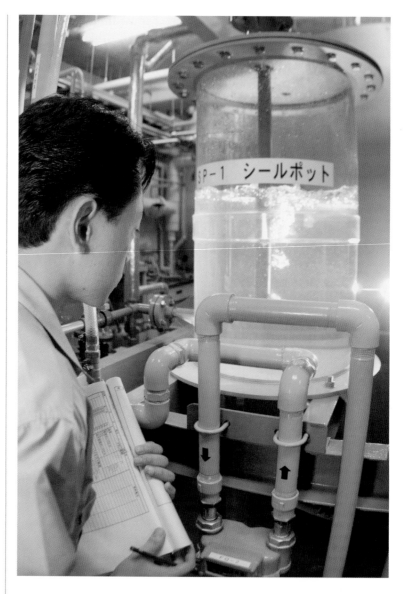

Impact of biogas

Biogas appears to offer many environmental benefits. It uses waste materials that would otherwise take up space in landfills or pollute the landscape to generate fuel. The fuel it creates is far less polluting than most fossil fuels. When methane burns, it produces carbon dioxide and water, so it does not cause the same degree of air pollution as fossil fuels. It does not produce the sludge that results from coal-burning emissions. Burning methane releases no ash and only small amounts of sulfur dioxide or nitrogen oxides and does not contribute to the formation of smog. Methane and carbon dioxide, the main components of biogas, are themselves pollutants, but burning the biogas prevents these pollutants from being released into the atmosphere.

On an economic level, biogas technology can save individual producers a great deal of money on power costs. For example, a dairy farm that implements biogas technology can save thousands of dollars every year on electricity, heating, and manure pit maintenance. On the other hand, installing the technology is very expensive; it can take several years to earn back the investment. Maintenance costs are also a factor. Some estimates predict that it would take a dairy farm more than five years to earn back the investment in a biogas operation, which is too long for most businesses to find financially acceptable.

Issues, challenges, and obstacles of biogas

Biogas technology is still being developed. It is difficult to persuade people to invest a great deal of money in equipment to collect and use biogas when they already have good equipment that uses fossil fuels. Few people know about biogas so there is not yet great demand for biogas appliances. China has used biogas from sewage fairly widely in the mid-twentieth century, on cooperative farms. There were successes, but the appliances have been difficult to maintain.

ETHANOL AND OTHER ALCOHOL FUELS

It is possible to use alcohol to power engines, either by itself or mixed with gasoline or other fuels. Ethanol is the most common of the alcohols that can be used to power engines. Ethanol is also known as ethyl alcohol and is the same kind of alcohol found in alcoholic beverages. It is clear and looks like water. but it is not the only one.

Methanol, or methyl alcohol, and butanol can also be used as fuel. Methanol is an alcohol made from fermentation of cellulose or from fossil fuels, particularly methane. It is used mainly as a fuel for race cars. Butanol is an alcohol made from fermenting plants. It can also be used as a fuel for internal combustion engines. Propanol is another kind of alcohol fuel. Methanol, butanol, and propanol all have the disadvantage of being toxic to humans and highly volatile (explosive). Ethanol is also volatile and toxic, but the toxicity level is lower, and so is considered more acceptable. Regardless of which one is used, alcohol combined with gasoline results in a fuel called gasohol.

An employee of Toshiba Corporation adds methanol liquid into the company developed prototype direct methanol fuel cell (DMFC) powered HDD-based digital audio player. The HDD player can run for approximately 60 hours on a single 10ml charge of pure methanol. ©Issei Kato/Reuters/Corbis.

Blends of gasoline and alcohol are often identified by abbreviations that combine the letter E with a number indicating the percentage of ethanol in the blend. For example, E10 contains 10 percent ethanol, E5 contains 5 percent ethanol, and E7 contains 7 percent ethanol.

How to make ethanol

Ethanol can be made from a large number of organic materials, including corn, wheat, grass, sugarcane, seaweed, cellulose left over from making paper, and nearly any other source of carbon. It can also be made from leftover petroleum feedstocks.

To make ethanol, a producer grinds up the feedstock, such as corn. This exposes the starch in the plant material. The ground-up corn is mixed with water and enzymes and heated to convert the starch to sugar. The producer adds yeast to the mix to help the sugars ferment into ethanol. The alcohol is then removed by a process called distillation: The producer boils the mixture so that

Do Not Drink the Ethanol

Humans long ago figured out how to make ethyl alcohol. It is fairly easy to do; any source of sugar will create the fermentation that results in drinkable alcohol. Ethanol producers, however, ruin their liquid for human consumption. First they add benzene to the ethanol to remove any water that might be lingering in it, which would impair its ability to function as a fuel. Drinking ethanol with benzene in it can damage the liver. Before the ethanol is sold, the producer "denatures" it by adding some poisonous substance to it. A popular choice for this poison is methanol, also known as methyl alcohol or wood alcohol, which is terribly toxic to humans.

the alcohol evaporates and then catches the alcohol in a container and cools it back into a liquid.

Sugarcane is the best source of ethanol because it naturally contains the sugars that ferment into alcohol. Scientists are working on better methods of making ethanol from cheaper biomass materials, such as wood and straw. It is harder to make ethanol from these substances because they do not release their sugars as easily as corn or sugarcane.

Current uses of ethanol and other alcohol fuels

Ethanol and other alcohols can be used to power motor vehicles instead of gasoline. In almost all cases the ethanol is mixed with gasoline. Gasoline-powered vehicles have no difficulty using gasoline that contains small amounts of ethanol. Generally this mix must contain at least 10 percent ethanol to qualify as gasohol. Gasohol is widely available in Denmark, Brazil, and the American Midwest. The state of Minnesota requires all gasoline sold there to contain at least 10 percent ethanol.

Increasing numbers of light trucks are sold as flexible fuel vehicles, capable of burning a variety of fuels, including mixes of gasoline and ethanol and other alternative fuels such as P-Series fuels. Vehicles that can run on pure ethanol are rare and require special engineering to function, which is why fuels for FFVs usually contain at least some gasoline.

One common ethanol blend is called E85, which contains 15 percent gasoline and 85 percent ethanol. Producers add this small amount of gasoline to the ethanol to make the vehicle start better in cold weather. E85 is generally priced at about the same level as gasoline.

Many scientists also hope that ethanol can be an important source of fuel for fuel cells in the future. Ethanol and methanol can both be used as fuels in fuel cells, though ethanol is a less efficient source than methanol. Fuel cells would use the energy stored and released by hydrogen.

Ethanol also has many other uses. It has a low melting point, so it can be added to liquids as an antifreeze. In addition, it can be added to gasoline as an anti-knocking agent. It can also be a safe replacement for MBTE, a fuel additive that has been found to present environmental problems.

Benefits and drawbacks of ethanol

Because ethanol can be made from so many different substances, it can be made nearly anywhere from nearly any raw material. Most ethanol is made from corn and sugarcane, but scientists have been investigating other sorts of biomass as a source of ethanol. Cellulose from grass or hay, cardboard, paper, farm wastes, and other waste products could potentially produce much more energy per source than is currently possible, with the side benefit of using up organic waste matter that would otherwise be thrown into landfills.

Ethanol is less flammable than gasoline and thus may be less of a fire hazard. When it does catch on fire, however, its flame and smoke are very hard to see, which presents another set of risks. Ethanol and other alcohol fuels dissolve in water, so water will put out alcohol fires, unlike gasoline fires, which require special fire extinguishers.

Ethanol will dissolve rubber and plastic, so pure ethanol cannot be used in unmodified gasoline engines. Also, ethanol's octane rating is higher than gasoline, which can require modifications to spark timing, carburetor jets, and starting systems. Gasohol does not present the same problems and can be used in ordinary vehicles without modification.

Environmental impact of ethanol

The environmental implications of making and using ethanol are the source of much debate. While burning ethanol has many environmental advantages over gasoline, particularly in reduced air pollution, the production of ethanol can be decidedly un-green.

Ethanol does not emit the same greenhouse gases that gasoline does. When it burns, it emits only carbon monoxide and water. Air quality improves quickly when ethanol replaces gasoline. Minnesota, which requires all its gasoline to contain 10 percent ethanol, has met Environmental Protection Agency (EPA) carbon monoxide targets partly because the ethanol has reduced the amount of gasoline burned.

Ethanol has the potential to reduce garbage in landfills. If ethanol can be made from waste paper or wood, that would supply a use for what has historically been a big source of trash. On the other hand, the process of creating ethanol from waste cellulose itself creates waste products that cannot be used.

Ethanol production does come with some environmental problems, however. Many experts contend that ethanol made from corn is actually worse for the environment than fossil fuels. This is because it can take more energy to raise the corn and make ethanol than the resulting ethanol can itself provide. Commercial farms use vast amounts of fossil fuels in planting, harvesting, and fertilizing their crops and making ethanol. In the United States the corn ethanol industry has been heavily subsidized (supported) by the government, which makes it inexpensive to manufacture ethanol from corn crops. If, however, the industry uses more energy to make ethanol than ethanol can provide, then ethanol is in fact not a workable alternative to gasoline.

Economic impact of ethanol

For states that produce corn, ethanol adds a great deal of value to local corn crops. For example, in Minnesota about 14 percent of the corn crop is made into ethanol. Exporting ethanol instead of raw corn doubles the value of the corn. Many midwestern states have subsidized ethanol production from corn since the 1970s, when Middle Eastern nations instituted an oil embargo in 1973. The U.S. federal government has guaranteed loans to build ethanol plants and since 1978 has made gasohol exempt from (free of) certain taxes.

For individual consumers, the cost of running a vehicle on gasohol is about the same as running it on gasoline, though that varies widely with the price of oil. As oil prices rose in the early 2000s, ethanol became comparatively cheaper. During this time, as it became apparent that biofuels were becoming widely accepted, production of ethanol increased very rapidly around the world. Ethanol appeared poised to become a giant and lucrative (money-making) industry.

Issues, challenges, and obstacles of ethanol

In areas where it is easy to make ethanol, such as Brazil, with its ample water and warm climate that makes it easy to grow sugarcane, ethanol is an entirely viable fuel. The nation powers its ethanol plants by burning bagasse, the sugarcane solids, which can generate enough power to have some left over. Hydroelectric power is also a good way of making ethanol without using fossil fuels.

The main source of debate about ethanol is whether or not making and using ethanol is actually more efficient than using straight fossil fuels. The problem is that producing ethanol consumes a great deal of energy. First, a farmer must grow the grain or sugarcane that provides the source of most ethanol, which takes up agricultural land and consumes water and fertilizers, many of them made from fossil fuels. The process of making and transporting ethanol consumes energy. Natural gas is a commonly used fuel in the distillation process, and it is itself a fossil fuel. Critics of ethanol have long insisted that making ethanol from corn costs more energy than the resulting ethanol can produce.

Critics also claim that corn-growing states in the United States have been emphasizing the importance of corn-based ethanol to get subsidies from the federal government that are far out of proportion to ethanol's value to the economy. Corn growers have exerted a great deal of political power, and agricultural states have used ample influence in national politics. Critics fear that ethanol producers will persuade the government to invest in their industry despite the fact that it may not have real environmental benefits.

P-SERIES FUELS

P-Series fuels are a new type of renewable fuel that use up an extremely common and little-valued resource: garbage. P-Series fuel is a blend of 35 percent natural gas liquids, 45 percent ethanol, and 20 percent methyltetrahydrofuan (MeTHF). The natural gas liquid is a substance called pentanes-plus, a liquid left over from the processing of natural gas, with butane added in winter months. MeTHF is made from biomass such as waste paper, food wastes, agricultural waste, or yard waste, and serves as a co-solvent (substance that turns another into liquid). The fuel is a colorless clear blend with octane between 89 and 93, the same octane as gasoline. It can be formulated for winter or summer use. It can be used alone or mixed with gasoline in a flexible fuel vehicle (FFV).

P-Series fuel was developed in the 1990s by Princeton University thermonuclear physicist Stephen Paul. He wanted to create a substitute for gasoline and thought that using garbage as fuel could work. He gave the fuel its name in honor of Princeton University. Paul and fellow investors have bought a sludge plant in New Jersey that they intend to use to make enough P-Series fuel to power about fifteen thousand vehicles.

Current uses of P-Series fuels

P-Series fuels are not currently widely used. They are still quite new, and no car manufacturer has yet produced a "P-Series-specific" FFV. If consumers begin buying these fuels, however, they could be a good substitute for gasoline.

Benefits and drawbacks of P-Series fuels

Using P-Series fuels has several benefits. It decreases the amount of petroleum used to power vehicles. It makes use of waste that would otherwise have to be placed in a landfill, incinerated, or transported to some other location. P-Series fuels are easy to use. Fueling an FFV with P-Series fuel is identical to fueling a vehicle with gasoline. There is no need to monitor fuels because gasoline and P-Series fuels will work mixed together, so a car owner can fuel up at ordinary gas stations or at P-Series pumps without thinking about which is which. This is especially useful when traveling to areas where P-Series fuels are unavailable.

But P-Series fuels cannot be used in vehicles designed to burn gasoline only. FFVs designed to burn methanol or ethanol can burn it, but ordinary cars cannot. P-Series fuels are slightly more efficient than gasoline, but in practice, mileage for vehicles using P-Series fuels is about 10 percent less per gallon than those using gasoline.

Environmental impact of P-Series fuels

The feedstock used to make MeTHF is chemically digested by the process of making it; as a result, the raw material is completely consumed and no emissions enter the air. Burning P-Series fuels in vehicles releases many fewer emissions than burning fossil fuels. In fact, P-Series fuels were added to the list of alternative fuels under the U.S. Energy Policy Act in 1999.

Economic impact of P-Series fuels

In 2003 P-Series fuels cost about $1.49 per gallon, which was then slightly lower than gasoline. Because P-Series fuels provide slightly less power than gasoline, the resulting operating cost is about the same for a P-Series-powered car and a gasoline-powered car. It is possible that as fossil fuels become more expensive, P-Series fuels will seem less expensive.

Manufacturers of P-Series fuels usually buy their natural gas liquids and ethanol in bulk from companies that produce those products. MeTHF is made by hydrolysis; the process basically involves mixing garbage with some acid and heat and agitating (mixing) it until it turns into liquid. The feedstock, or raw material, for MeTHF actually has a negative cost, because it is made from materials that would otherwise cost a city money to dispose of. As a result, it is fairly easy for a P-Series plant to recoup (get back) its investments and become profitable. Small P-Series plants are viable because it is not very expensive for them to operate. This makes it possible for many small P-Series plants to be distributed throughout a geographic area. This distribution would have the added advantage of preventing any one location from becoming the region's dumping ground.

Issues, challenges, and obstacles of P-Series fuels

P-Series fuels are still very new and appear to be unproven. Producers of the fuel have a hard time finding investment capital for their enterprises because banks or companies investing in the project want to be sure they can collect a return on their money. The developer of the fuel insists that it burns cleanly and that it will in fact be inexpensive to make. Without a provable record, however, it is difficult to persuade investors that this is true.

■■■

For More Information

Books

Carter, Dan M., and Jon Halle. *How to Make Biodiesel.* Winslow, Bucks, UK: Low-Impact Living Initiative (Lili), 2005.

Pahl, Greg. *Biodiesel: Growing a New Energy Economy.* Brattleboro, VT: Chelsea Green Publishing Company, 2005.

Tickell, Joshua. *From the Fryer to the Fuel Tank: The Complete Guide to Using Vegetable Oil as an Alternative Fuel.* Covington, LA: Tickell Energy Consultants, 2000.

Periodicals

Anderson, Heidi. "Environmental Drawbacks of Renewable Energy—Are They Real or Exaggerated?" *Environmental Science and Engineering* (January 2001).

Parfit, Michael. "Future Power: Where Will the World Get Its Next Energy Fix?" *National Geographic* (August 2005): 2–31.

"Stirrings in the Corn Fields." *The Economist* (May 12, 2005).

Web sites

"Alternative Fuels Data Center." U.S. Department of Energy: Energy Efficiency and Renewable Energy. http://www.eere.energy.gov/afdc/altfuel/p-series.html (accessed on July 11, 2005).

American Council for an Energy-Efficient Economy. http://aceee.org/ (accessed on July 27, 2005).

Biodiesel Community. http://www.biodieselcommunity.org/ (accessed on July 27, 2005).

"Bioenergy." Natural Resources Canada. http://www.canren.gc.ca/tech_appl/index.asp?CaId=2&PgId=62 (accessed on July 29, 2005).

"Biofuels." Journey to Forever. http://journeytoforever.org/biofuel.html (accessed on July 13, 2005).

"Biogas Study." Schatz Energy Research Center. http://www.humboldt.edu/~serc/biogas.html (accessed on July 15, 2005).

"A Complete Guide to Composting." Compost Guide. http://www.compostguide.com/ (accessed on July 25, 2005).

"Ethanol: Fuel for Clean Air." Minnesota Department of Agriculture. http://www.mda.state.mn.us/ethanol/ (accessed on July 14, 2005).

"Fueleconomy.gov." United States Department of Energy. http://www.fueleconomy.gov/feg/ (accessed on July 27, 2005).

Geothermal Energy

INTRODUCTION: WHAT IS GEOTHERMAL ENERGY?

Geothermal energy is energy created by the heat of the Earth. Under the Earth's crust lies a layer of thick, hot rock with occasional pockets of water. This water sometimes seeps up to the surface in the form of hot springs. Even where the water does not travel naturally to the Earth's surface, it is sometimes possible to reach it by drilling. This hot water can be used as a virtually free source of energy, either directly as hot water, steam, or heat or as a means of generating power. Geothermal energy is nonpolluting, inexpensive, and in most cases renewable, which makes it a promising source of power for the future.

The word *geothermal* comes from two Latin words, *geo*, meaning "earth," and *thermal*, meaning "heat." So the word *geothermal* means "heat from the earth." In most cases, the geothermal resource that people want is water that has been trapped within the Earth, where it becomes very hot.

Types of geothermal energy

There are two main types of geothermal energy. The energy can be used directly, as heat or hot water, or it can be a means of generating electricity.

Naturally hot water has been recognized as a resource for thousands of years. People have used hot springs for bathing, for medical treatments, and as heating for their buildings. The hot water can also be used in agriculture, aquaculture, industry, and other applications.

Geothermal power can also generate electricity. Geothermally generated electricity is becoming increasingly important. In 1999 over 8,000 megawatts of electricity were produced by about 250

Words to Know

Aquaculture The formal cultivation of fish or other aquatic life forms.

Balneology The science of baths, especially for therapeutic use.

Core The center, innermost layer of the Earth.

Crust The outermost layer of the Earth.

Geothermal reservoir A pocket of hot water contained within the Earth's mantle.

Lava Molten rock contained within the Earth that emerges from cracks in the Earth's crust, such as volcanoes.

Magma Liquid rock within the mantle.

Mantle The middle layer of the Earth between the inner core and the outer crust.

Turbine A device that uses the movement of a liquid or gas to spin a machine that produces electricity.

geothermal power plants around the world, located in twenty-two different countries. Most of these power plants are located in developing nations. However, that same year the United States produced nearly 3,000 megawatts of geothermal electricity, more than twice the amount of power generated by wind and solar power. Ten percent of the electricity in Nevada and 6 percent of the electricity in Utah came from geothermal power plants.

Historical overview: notable discoveries and the people who made them

Knowledge of geothermal energy is very old. Ancient Chinese and Japanese people bathed in hot springs and used the water for cooking. Ancient Romans used the water from hot springs as a medicine for skin diseases, and the buildings in ancient Pompeii were heated with hot water that ran under them. Native Americans settled near hot springs more than 10,000 years ago. During the Middle Ages in Europe people traveled to towns in Germany and France that had built spas, or health resorts, around natural hot springs.

Discoveries of the 1800s

During the 1800s European settlers moved westward across the North American continent. They noted the existence of hot springs and settled near them. In 1807 John Colter (1774–1813) is believed to have found hot springs in what is now Yellowstone National Park. That same year the city of Hot Springs, Arkansas, was founded. By 1830 Asa Thompson of Hot Springs was selling visitors the right to sit in a wooden tub fed by a hot spring; the price was $1 per person. The hot springs area in Arkansas was declared a national park in 1921.

Bath, England, with its natural hot springs, is the site of an elaborate Roman public bath built in the first century C.E. © *Bob Krist/ Corbis.*

In 1847 William Bell Elliot, a member of John Fremont's California survey group, found a valley full of steaming hot springs that he described as resembling the gates of hell. He named the area "the Geysers" (though it did not actually contain geysers). The region was located north of San Francisco. Within five years the area had been developed into a resort spa that was visited by famous people such as the author Mark Twain and the presidents Theodore Roosevelt and Ulysses S. Grant. Ten years later Sam Brannan built a $500,000 resort southeast of the Geysers called

Hot Springs Baths

The ancient Greeks and Romans knew of a number of natural hot springs, many of them located near volcanoes. The oldest known hot springs bath still in existence is located in Merano, Italy. People are believed to have used it five thousand years ago. Bath, England, has long been famous for its natural hot springs. The waters at Bath are about 120°F (48°C) and contain numerous minerals, including calcium and magnesium. Ancient Celts are believed to have bathed in the springs as early as 800 BCE (before the common era). The Romans built bath houses around the springs nearly two thousand years ago; the town became a major tourist resort starting around the time of Queen Elizabeth I (1533–1603), who went there often to bathe.

Germany is full of natural hot springs, many of which long ago became the sites of baths. Ancient Romans built baths on these springs. Like Bath in England, Germany's bath towns became extremely popular with the rich in the nineteenth century. Towns such as Bad Cannstatt and Baden-Baden grew rich from their well-to-do visitors who came to bathe, be massaged, drink the waters, and indulge themselves in other entertainments. These baths are still popular today and have been supplemented with modern healing treatments such as shiatsu massage and with trendy shopping facilities. Japanese baths are likewise famous around the world. Hot springs resorts, called *onsen*, attract millions of visitors who come to soak in the waters. The waters often contain particular minerals that are said to have specific effects on physical and mental health. Some baths have facilities for drinking the water or inhaling the steam.

Calistoga, which resembled European resorts with racetracks, bathhouses, a hotel, and a skating pavilion.

Americans began experimenting with large-scale geothermal heating in 1864, with the construction of the Hot Lake Hotel in La Grande, Oregon. In 1892 the city of Boise, Idaho, built a geothermal district heating system that piped hot water from a geothermal reservoir to the buildings in town.

Beginning of geothermal electricity

The first geothermal electrical power plant was built in Larderello, in Tuscany, Italy, in 1904. Larderello is a geologically active area that was used in Roman times as a hot springs resort. This made it ideal as a site for experimenting with geothermal energy. The first plant lit up five light bulbs, using the steam that came from cracks in the ground. In 1911 a larger plant opened in the area, and it was the only geothermal power plant in the world until after World War II. The plant at Lardarello was destroyed during World War II, but it was quickly rebuilt. Engineers from New Zealand and other countries went to visit the Larderello plant to learn how it was built and also noted the enthusiasm that the Italian engineers had for their plant. Lardarello's plant still produces enough power for one million households in Italy, nearly ten percent of the total geothermal power produced in the world.

In 1921 John D. Grant drilled a well at the Geysers with the hope of using its steam to generate electricity. The next year he built the first geothermal power plant in the United States. His

People bathing in Blue Lagoon near Grindavik, Iceland © *Hans Strand/ Corbis.*

Hot Springs Monkeys

Humans are not the only creatures to have noticed and taken advantage of natural hot springs. Japanese macaques, also known as snow monkeys, are large monkeys that live in northern Japan. The Japanese winter is cold and snowy, but the macaques have learned a trick that helps them keep warm: They sit in natural hot springs that come up from the ground. The monkeys got so enthusiastic about hot springs that the prefecture (governmental district) of Nagano decided to build them their own hot springs and feeding stations to keep them away from human hot tubs and spas.

power plant generated enough power to power the lights at the Geysers resort. However, geothermal power at the time cost more than other sources of power to produce, so this effort was soon abandoned.

Through the 1920s people continued to drill experimental wells in Oregon and California, hoping to take advantage of the heat within the Earth. In 1927 the Pioneer Development Company drilled some wells in Imperial Valley, California. In 1930 gardeners in Boise, Idaho, opened the first geothermally heated commercial greenhouse, using the water from a 1,000-foot (305-meter) well. That year Charlie Lieb of Klamath Falls, Oregon, built the first downhole heat exchanger, which he used to heat his home. In 1940 the Moana neighborhood of Reno, Nevada, began using geothermal heat for residential heating. Eight years later the first groundwater heat pumps went into use in Ohio and Oregon.

The first flashed steam geothermal power plants, which depressurized hot water to produce steam, were built in the late 1940s. In 1960 the United States' first large-scale geothermal power plant began operation, at the same site and with the same name (the Geysers) as the earlier spa. Its first turbine produced 11 megawatts of net power. As of the early 2000s the Geysers was the largest geothermal plant in the world.

Governmental encouragement

In the 1970s the United States and other nations created several agencies and passed laws to encourage the development of geothermal energy. In 1970 the United States passed the Geothermal Steam Act, which gave the Secretary of the Interior the authority to use public lands for environmentally sound geothermal exploration and development. The Geothermal Resources Council was formed to make it easier to develop geothermal resources worldwide. The Geothermal Energy Association, founded in 1972, was created by several companies around the world to develop geothermal electricity generation and direct heat technology. The 1974 Geothermal Energy Research, Development and Demonstration Act instituted a geothermal loan guaranty program, which gave investment security to companies attempting to create technologies to use geothermal energy. In 1975 the Geo-Heat Center was formed at the Oregon Institute of Technology; the institute began using runoff from its geothermal heating system to heat water used to raise freshwater prawns.

The first geothermal food processing and crop drying plant was opened in Brady Hot Springs, Nevada, in 1978. It received $3.5 million from the Geothermal Loan Guaranty Program. That year the United States Department of Energy opened a facility in Fenton Hill, New Mexico, to test "hot dry rock" energy generation, a process in which water is pumped into an area of hot rock, becomes superheated, and then is pumped back to the surface so that the heat can be siphoned off. This facility managed to generate some electricity two years later.

Between 1979 and 1982 the Department of Energy sponsored development of a geothermal electrical power plant in Imperial Valley, California, as well as research into direct uses of geothermal energy for heating and agriculture. The first flashed steam plant in the United States was built in Brawley, California, in 1980. In 1981 a binary power plant was built in California's Imperial Valley. The plant was so successful that Ormat, the company that built it, paid off its loan within one year. By 1984 there were geothermal power plants in Hawaii, Nevada, and in the Salton Sea in California.

In 1989 the first hybrid geothermal power plant opened in Pleasant Bayou, Louisiana. It used both geothermal heat and methane to create electricity. During the 1990s several geothermal power plants went into operation in the Pacific Northwest, Nevada, and Hawaii. In 1994 the United States Department of Energy created two programs to increase the use of geothermal power generation and heat pumps in an effort to reduce greenhouse gas emissions.

A geodesic dome at the geothermal power plant in Nesjavellir, Iceland. The plant sends heated water to the city of Reykjavik. © Roger Ressmeyer/Corbis.

In 2000 the U.S. Department of Energy created its GeoPowering the West initiative, which funded twenty-one partnerships with private companies to develop geothermal energy in the western United States. Several groups in the western states spent the early 2000s working to identify barriers to geothermal development and to create ways to make geothermal energy more commonly used.

How geothermal energy works

Geothermal energy uses the heat of the Earth to produce electricity and heat. This form of power works because the inside of the Earth is much hotter than the surface.

The structure of the Earth

The Earth consists of several layers of matter. The outer layer, called the crust, is the surface where people live and plants grow. It is composed of aluminum, silicon, oxygen, iron, and other minerals. Below the crust is a layer called the mantle, a thick layer of rock and oxides that comprises about 82 percent of the Earth's total volume. It is made up mostly of

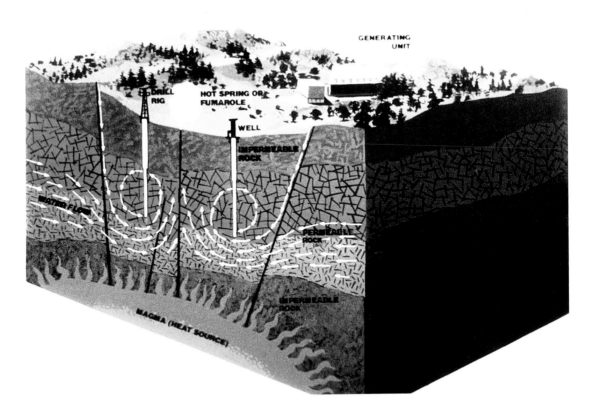

GENERATING
UNIT

DRILL
RIG

HOT SPRING OR
FUMAROLE

WELL

IMPERMEABLE
ROCK

WATER FLOW

PERMEABLE
ROCK

IMPERMEABLE
ROCK

MAGMA (HEAT SOURCE)

Cutaway drawing of the Earth, showing source of geothermal energy. U.S. Department of Energy, Washington D.C.

peridotite, a kind of rock containing iron, magnesium, oxygen, and silicon. The mantle is mostly solid but can also flow like a liquid when it is under pressure. The top layer of the mantle consists of hot liquid rock called magma. The crust floats on top of this liquid rock.

At the center of the Earth is the core, a chunk of extremely hot iron and nickel. The core itself consists of two layers, the outer core, which is liquid, and the inner core, which is solid because of the tremendous pressure it experiences. The center of the core is about 4,000 miles (6,400 kilometers) from the surface of the Earth.

Water heated underground

The Earth's temperature increases about 41.7 °F (5.4 °C) for every 328 feet (about 100 meters) traveling from the surface to the core. About 10,000 feet (3,048 meters) below the surface, temperatures are hot enough to boil water. The inner core may be over 9,000 °F (4,982 °C). This heat constantly travels upward toward the surface, heating the mantle, which carries heat toward the crust. Similar to the curved pieces of peeled skin from an orange, Earth's outermost layers are cut and fractured into pieces or sections called plates. Like the inner and outer sides of an orange peel, these plates have distinct sections, an inner side and an outer crust. Each plate (also called a lithospheric plate) moves over a hotter, denser—but in many ways more fluid-like (molten)—region of Earth's interior termed the asthenosphere (a portion of Earth's mantle). The visible continents such as North and South America are actually an outer crust of the lithospheric plates upon which they ride, shifting slowly over time as a result of forces, including differences in temperature, which help move or drive the plates. The theory that describes this motion is perhaps the most important in all of geology (the study of Earth's structure) and is called plate tectonics (the theory of plate structure and movements). Although plates move very slowly (in many cases, just inches per year) they are, of course, very heavy and so their rubbing, sliding, slipping, collisions, and bending causes earthquakes.

Where the edges of plates overlap, volcanoes may form. Depending on the materials that compose them, one plate may drive under another (subduction) or both plates may drive skyward to form mountain chains. Hot magma from Earth's molten inner layers (or from pieces of plate being destroyed during subduction) can carve tunnels, chambers, and channels in the plate and crust and so allow hot magma to reach the surface of the plate (even if it is under the ocean). When magma reaches the surface and flows from a volcano it becomes known as lava. Volcanoes can also form over areas in plates away from the edges (especially thinner areas of plates under the oceans) called "hot spots" where molten material from Earth's mantle pushes upward.

The rock underground is full of cracks and small pockets, and these can fill with water. Water that gets trapped in underground caves will get very hot, even hotter than boiling temperature, but it cannot boil because there is no place for steam to escape into the air. This water sometimes finds its way to the surface in the form of hot springs. Most of the hot water stays underground in pockets called geothermal reservoirs.

Making use of geothermal energy

There are several ways to make use of geothermal energy. The most basic is simply to use the water as hot water when it comes out of the ground. The water can be channeled to different places as heat, for heating homes, or for cooking.

Engineers can drill down into the ground to reach geothermal reservoirs and then use the hot water, steam, or heat to power generators to make electricity. Scientists have developed techniques to find geothermal water. When they find reservoirs, they drill production wells down into them. The hot water or steam travels up the well to the surface, where it can be collected and harnessed for various uses.

A fumarole bubbles at a Pacific Gas and Electric Company geothermal power plant in California. © *Roger Ressmeyer/Corbis.*

The Ring of Fire and other hot spots

The Pacific Ocean is one of the most geologically active areas in the world. The land that borders the Pacific is sometimes known as the Ring of Fire because of the volcanic activity that occurs there. New Zealand, Japan, the Philippines, Hawaii, Alaska, California, and other places in the area experience a great deal of tectonic shifting, as pieces of the Earth's crust move around and crash into one another. All of these areas also have active volcanoes.

There are active volcanoes in many other places. Iceland has so much volcanic activity that it derives much of its power from

Geysers, Hot Springs, Mudpots, and Fumaroles

Magma heats water trapped or flowing underground. Hot springs are places where hot water rises up from the Earth on a regular basis. Geysers are explosive hot springs where hot water periodically shoots out of a hole in the ground. Fumaroles are openings near volcanoes that emit steam and sulfurous gases. They can look like holes or cracks in the ground and may stay in the same spot for centuries or come and go within weeks. Mudpots are fumaroles or hot springs that form in areas with small amounts of water. The water bubbles up to the surface and creates a crater filled with boiling mud.

geothermal sources. Kenya, Turkey, Italy, and Zambia all have enough geothermal energy to make profitable use of it. Because of the nature of current geothermal technology, these geologically active areas are also the main sites of geothermal power.

Current and future technology

Geothermal energy technologies are used in the generation of electricity and in direct uses of the hot water. There is room for development of new technologies in both categories.

Geothermal power plants

One of the most important uses of geothermal energy is to generate electricity. In geothermal power plants, hot water drawn from geothermal reservoirs through production wells spins turbine generators, which produce electricity. The used water is injected back into the reservoir through another well called an injection well. This water gets hot again and helps maintain the pressure within the reservoir. If all the water were removed and not replenished, the reservoir would eventually cool off and run out of water, making it useless. Groundwater must be very hot in order to generate electricity. Water colder than 250 °F (121 °C) is currently not usable for power.

There are three main types of geothermal power plants.

- Flashed power plants have reservoirs with water between 300 and 700 °F (148 and 371 °C). This water comes up from the well and is flashed (turned quickly) into steam, which powers a turbine.
- Binary power plants have reservoirs with water between 250 and 360 °F (121 and 182 °C), which is not quite hot enough

Sulfur extraction from Geysers geothermal steam power plant operations, Sonoma County, California. ©*Gerald & Buff Corsi/Visuals Unlimited. Reproduced by permission.*

to generate enough steam to power a turbine. These plants use the heat from the water to heat another liquid with a lower boiling temperature, called a binary liquid. The binary liquid boils and produces steam to spin a turbine.

- Dry steam power plants have reservoirs that produce steam but not water. The steam is piped directly into the plant, where it spins a turbine.

There are also hybrid power plants that combine geothermal heat with other sources of energy, such as methane. All types of geothermal power plant have no emissions and can produce a large amount of power. Geothermal power is especially appealing because it is possible to have power plants of almost any size, from tiny 100 kilowatt plants to much larger 100 megawatt plants that are connected to national power grids. They can operate twenty-four hours a day every day of the year, but they can also vary operation according to demand.

Direct uses of geothermal energy

Hot water is useful in and of itself. Some common uses of geothermal water include:

- Using hot springs for bathing. This is called balneology.
- Growing plants in winter greenhouses.
- Heating the ground in which outdoor crops are growing to prevent it from freezing.
- Growing fish and shellfish for commercial purposes.
- In industry, such as pasteurizing milk or washing wool.
- Heating buildings or cities through underground channels. Reykjavik, Iceland, has the world's largest geothermal district heating system.
- Piping water under streets and sidewalks to keep them from freezing.
- Geothermal heat pumps that use the heat from just a few feet below the Earth's surface instead of heat from geothermal reservoirs. These can heat or cool homes anywhere, not just in areas with geothermic activity.

Every use of geothermal water as hot water saves energy. Heating water takes a great deal of power, and every gallon that does not have to be heated can save oil, coal, wood, or other heating fuels.

Direct uses of geothermal energy provide about 10,000 thermal megawatts of energy in thirty-five countries around the world. This does not include the use of geothermal waters for bathing by individuals who have not developed the resources for commercial use. In the United States in the late 1990s there were eighteen district heating systems, twenty-eight fish farms, thirty-eight greenhouse establishments, twelve factories, and more than two hundred spas using geothermal waters.

Developing technology

Scientists are working to create technology that will make geothermal energy more accessible to people everywhere, not just to those living in areas with shallow geothermal reservoirs. The entire planet has heat beneath its surface, but not all places have hot water. Deeper drilling techniques could make more areas of heat and steam accessible. Scientists would love to take advantage of the heat from magma in the mantle, but there is not yet a workable technology to do this.

Engineers are working to develop technology that would make hot dry rocks (HDR) 3 to 6 miles (5 to 10 kilometers) below the surface usable for power. Techniques include piping water down to the hot rock to create steam. Teams in the United Kingdom, Australia, France, Switzerland, and Germany are working on HDR technology as of the early 2000s. It remains to be seen if they can devise a method of producing power that is worth its cost.

Benefits of geothermal energy

There are many benefits to using geothermal power. It is clean and nonpolluting. It does not require the consumption of fossil fuels, so it reduces dependence on foreign or domestic oil, and it reduces harmful emissions from burning these fuels. Geothermal plants do not destroy large tracts of land. They are efficient: A geothermal plant usually can produce more power than a fossil fuel-burning plant of the same size.

Geothermal plants are also very reliable. Because they do not depend on external fuel sources, they can run twenty-four hours a day, every day of the year. This is not always possible with power plants that burn coal or oil, which must be transported from distant locations. Geothermal plants are not vulnerable to weather, natural disasters, strikes, political disturbances, or other events that can disrupt fuel supplies.

Geothermal plants, on many levels, are flexible. It is possible to build them of modular components and to add or adapt components as the need arises. This is usually not possible with fossil fuel-burning plants. Geothermal power plants are especially valuable in areas with small power grids or in cases where a power grid is in the process of expanding. Flexible geothermal plants can provide backup power while the rest of the grid is installed.

Geothermal energy is generally sustainable and renewable. The Earth generates heat constantly. Rainfall and snowmelt continuously replenish reservoirs, and returning used water to the underground reservoir maintains its pressure and heat so that the reservoir can be used for an indefinite period of time.

Drawbacks of geothermal energy

The major limitation of geothermal power is that it can only be implemented in areas where there is a ready supply of hot water underground. This limits its use to geologically active areas such as California, Iceland, Japan and the rest of the Pacific Rim, and other areas with a thin crust, an active mantle, and pockets of subterranean hot water.

Only the hottest water can be used to generate electricity. Some places have naturally heated groundwater that is not hot enough to produce the steam needed to turn turbines. That water is still usable for other purposes but not as a workable power source.

It is possible to deplete a reservoir. If a geothermal reservoir runs out of water or grows too cool, it ceases to be useful, though this depletion can take decades or even centuries. For this reason some experts claim that geothermal energy is not actually a renewable resource.

In the early 2000s there are few areas with enough readily accessible geothermal water to produce electricity at a price that can compete with other sources of power. This may change as technology improves and other geothermal sources become usable or as the price of fossil fuels increases.

Environmental impact of geothermal energy

Like solar power and wind power, geothermal energy is clean. Geothermal power plants do not have to burn fuels so they do not produce emissions of greenhouse gases or other pollutants, which means they do not contribute to smog or global warming. They do emit very small amounts of carbon dioxide, about four percent of the amount emitted by burning fossil fuels. Binary plants produce no emissions at all. Areas that have geothermal power plants tend to have much better air quality than those with fossil fuel-burning power plants.

A geothermal power plant can be small compared to other types of power plants, so it is not as disruptive to the landscape. It can be built right next to its geothermal well. There is no need to build dams, dig mines, cut down trees, or dispose of wastes, which are necessary with other common forms of power. It is actually possible to build geothermal power plants in the middle of farmland or forests without damaging the surrounding plants and animals.

Even in areas where geothermal energy is not powerful enough to create electricity, people can still make use of local hot water for heating and bathing. This means they do not have to use electricity from other sources to heat their water, which can help save money and fuel.

There are some minor environmental drawbacks to using geothermal resources. Geothermal reservoirs sometimes contain hydrogen sulfide gas, which smells like rotten eggs and can be toxic at high concentrations. Geothermal power plants use scrubbers to remove this gas from emissions. Geothermal water also contains a high concentration of minerals, so geothermal wells must contain several layers of pipe and casings to prevent geothermal water from mixing with ordinary groundwater. Because geothermal power plants re-inject their used geothermal water back into the underground reservoir, in most cases the geothermal water never gets near groundwater and cannot harm aquatic plants and animals.

The areas around geothermal power plants experience increased activity, such as small earthquakes, and there is a danger of landslides. Federal laws in the United States prohibit the construction of geothermal power plants in national parks, such as Yellowstone. However, the environmental problems associated with using geothermal energy are generally far less serious than those caused by using fossil fuels.

Economic impact of geothermal energy

Geothermal power is produced locally, in the same area in which it is used. This means that states or nations do not have to pay other countries for fuel, as most countries do with fossil fuels. All economic benefits from a geothermal power plant remain in the area that produces the energy.

Using geothermal water saves money on other fuels, either to create electricity or to heat water. In the late 1990s, worldwide use of geothermal energy saved the equivalent of 830 million gallons of oil or 4.4 million tons of coal. This amount could be increased considerably if the use of geothermal energy were expanded.

Societal impact of geothermal energy

Much of the world's geothermal energy is used by developing nations that cannot afford to use fossil fuels for power and that may not have other sources of energy. Thailand, Indonesia, the Philippines, and the Azores have all been making use of geothermally generated electricity since the late 1980s or early 1990s.

Geothermal energy is a good, nonpolluting way for developing nations to build their infrastructures without destroying the landscape or polluting their air and water. The power produced by geothermal energy can raise standards of living in remote areas that are too far from other power sources. Because the energy is inexpensive, nations may be able to use the energy generated during off-peak hours for regional development projects, such as pumping water for irrigation. Local communities can be in complete control of their source of power, making them less dependent on their own government or foreign aid.

Many developing nations have created energy policies that emphasize using local resources for power, encouraging local private investment in energy, and expanding power into rural areas. Geothermal energy is very compatible with these goals because it can be locally run, and the resulting power used by the local community.

Barriers to implementation or acceptance

Many areas have geothermal potential that has not been tapped. During the twentieth century fossil fuels were a cheap and established source of power, and few areas have any incentive to spend the money to build geothermal power plants. People have not yet become aware of the many potential uses of geothermal water, so they are not taking full advantage of it. For example, in some areas naturally hot water is used for purposes that ordinary water could fulfill, such as irrigation of crops and municipal water supplies.

Many nations, both developing and advanced, have conducted initial investigations into their own geothermal potential. They have identified numerous geothermal reservoirs that could be used directly or converted into electricity, but they have not pursued the deep drilling needed both to confirm the reservoirs' potential and to exploit them. Geothermal energy does require large capital investments in its initial stages, and this investment comes with some risk. This initial investment deters (holds back) many governments and companies, as does the fact that it can take several years to achieve a return on the investment of building a geothermal power plant. Fossil fuel power plants earn back their investments much more quickly. It is also known, however, that geothermal power plants have long-term economic benefits, including low operating costs and long-term profits.

One major difficulty with developing geothermal power is access to land. Because geothermal power plants can only be built on or near geothermal reservoirs, power companies must be able to buy or lease this land.

In the early 2000s, most cities and buildings have been designed around fossil fuels and other more traditional sources of energy. Oil, gas, and coal companies do not want to see fossil fueled power plants closed because that would cause them to lose customers. Utilities do not want to have to rebuild existing power plants to convert them to geothermal power because that would be very expensive. In many countries the government grants monopolies to utility companies that make it possible for those companies to provide power at all times regardless of price fluctuations, but also make it impossible for alternative energy suppliers to compete in an open market.

AGRICULTURAL APPLICATIONS

Geothermal water is very useful in agriculture. Agricultural applications make direct use of geothermal water, using it to heat and water plants, to warm greenhouses, or to dry crops.

In agriculture, geothermal water is used mainly as a source of heat and moisture. Irrigation pipes can bring hot water to cold ground, making it possible to grow crops that would otherwise die. It can also be piped into greenhouses to keep them warm and to maintain humidity. As with most other uses of geothermal energy, geothermal agriculture is only practical in areas that have geothermal resources. It is possible in agriculture, however, to use geothermal water that is much too cold for power generation or even home heating. Only a few nations have thus far made much use of geothermal heat for agricultural purposes. They include the United States, Kenya, Greece, Guatemala, Israel, and Mexico.

Current uses of geothermal energy in agriculture

The main agricultural uses of geothermal water include heating and watering open fields, warming and humidifying greenhouses, and drying crops.

Open field agriculture

Geothermal water can be used to keep the soil in open fields at a steady warm temperature. Farmers run irrigation pipes under the soil to provide both water and heat to the crops. Cool-weather root crops and rapidly growing trees grow faster and more abundantly if the soil temperature is kept at about 70 °F (21 °C). Using geothermal water for irrigation extends the growing season and keeps plants from being damaged by low air temperatures.

Geothermal water can also sterilize soil to kill pests, fungus, and diseases that can harm crops. Sterilization requires very hot water so that the steam can be applied directly to the soil. The farmers either heat the soil from pipes underneath it, or they apply the steam above the soil and cover it with a plastic sheet to keep the heat inside.

Greenhouses

Greenhouses are buildings with clear plastic or glass walls and ceilings that trap solar heat to create a controlled atmosphere for growing plants. Greenhouses often benefit from another source of heat during the winter months. Heating greenhouses with geothermal water helps maintain a constant temperature, resulting in a more reliable crop and faster-growing plants. The water in the pipes can be released into the air inside the greenhouse, raising humidity if necessary.

There are several techniques used to heat greenhouses with geothermal water. These include plastic tubes, finned pipes, finned coils, soil heaters, or unit heaters. These parts can be combined according to water temperature and the preferences of the grower and the plants. For example, a grower producing roses would want to create a heating system with good air circulation and low humidity. A grower producing tropical plants could adjust the system to create high humidity and high soil temperatures. Chinese shiitake mushroom growers in Fujian province use geothermal heat in a greenhouse to speed production time.

Two large greenhouses at the La Carrindanga Project in Bahia Blanca, Argentina, have been using geothermal pipes to heat their facilities. These greenhouses have sliding glass side panels that can open and close to regulate humidity and heat, and misting systems to water plants and maintain moisture in the air. The geothermal water runs through pipes buried just beneath the surface of the soil, where the heat from the water easily reaches plant roots. Boxes containing dirt and seeds can sit on top of these pipes so that they receive heat from below. The beds grow vegetables, flowers, and indoor and outdoor plants from seeds and cuttings. Bahia Blanca has an unreliable climate and is not a very good location for outdoor agriculture, but its geothermally heated greenhouses are very productive and reliable.

Drying crops

The heat from geothermal water can also be used to dry crops and timber. For example, since the mid-1980s the Broadlands Lucerne Company in New Zealand has been using geothermal steam to dry alfalfa.

Hot water from below the ground is piped into the greenhouses, which are used for growing tomatoes. Steam is rising from the warm waters. *Martin Bond/Photo Researchers, Inc.*

Benefits and drawbacks of agricultural applications

Geothermally heated greenhouses are especially useful in marginal areas where the climate is unreliable. They make plant and vegetable production more efficient, and they reduce the time it takes seeds to germinate and grow to maturity. In addition, they make it possible to grow crops in the off-season, when such plants ordinarily would not grow and when they can be sold for higher prices. Farmers can grow plants under denser and more controlled conditions. They lose fewer plants and can make more precise commitments to buyers for future deliveries of crops.

However, geothermal water is not available everywhere. Not every farming operation can make use of geothermal resources because either there are none in the region or they are too difficult to reach. Installing equipment to pipe geothermal water into a farm can be expensive and time-consuming.

Impact of agricultural applications

Using geothermal water to enhance agriculture causes few environmental problems. It does not pollute the land because only water is emitted, although if the water is contaminated with heavy metals, such as mercury, this could cause a public health concern. The use of geothermal water could potentially result in farms being constructed in areas that would otherwise not be suitable for agriculture, which could destroy natural landscape and animal habitat.

Economically, using geothermal water in agriculture can be quite inexpensive. If geothermal wells already exist, then the farmers need invest only in steel or plastic pipes to transport the steam or hot water to the field, greenhouse, or drying facility. In many places the hot water is quite shallow and inexpensive to reach.

Despite this comparative lack of expense, even this level of equipment is too expensive for many individuals and businesses. There are many regions that have geothermal resources that could be used for agriculture that have not yet been able to take advantage of them. For example, the Oserian Development Company on the shores of Lake Naivasha, Kenya, grows flowers for market. It has considered using hot water from the Olkaria Geothermal field to sterilize the soil. As of the early 2000s this plan had not been implemented because of the cost.

AQUACULTURAL APPLICATIONS

Aquaculture is the raising of fish and other aquatic animals in a controlled environment—basically, it is the farming of fish, shellfish, and other freshwater or marine (saltwater) creatures. Using geothermal water in aquaculture helps keep water temperatures consistent, which increases survival rates and makes the creatures grow faster.

Low-temperature geothermal resources that are not hot enough to produce electricity are very useful to fish farmers. Animals grown in water of the proper temperature grow faster and larger than those in cold water or water with fluctuating temperatures. They are also more resistant to disease and die less frequently.

Fish farmers with access to geothermal water can use it to regulate the temperatures of their fish ponds. Though the mechanism to accomplish this can be complicated, basically what happens is that the fish farmer opens valves to allow geothermal water to flow into the fish ponds until they reach the desired temperature. The valves are then closed to prevent the water from getting too hot. The mechanism is similar to adding hot water to a bathtub to bring the temperature to the desired level.

Water flow can be adjusted throughout the year to account for air temperatures. Most ponds contain some mechanism to circulate the water and keep it all at an even temperature. Aquaculture operations usually have several ponds, which are kept small enough to be heated or cooled easily.

Current uses of aquacultural applications

Geothermal water has played a role in aquaculture for more than thirty years. In the 1970s the Oregon Institute of Technology began using runoff from the school's geothermal heating system to heat water used to raise freshwater prawns. In Arizona, fish farmers use geothermal waters between 80 and 105 °F (26 and 41 °C) to raise bass, catfish, and tilapia. The Salton Sea and Imperial Valley areas in southern California are home to about fifteen aquaculture operations. These fish farms produce about ten million pounds of fish every year, mostly catfish, striped bass, and tilapia, which are almost all sold in California.

People in other nations have also taken advantage of geothermal water for aquaculture. There are geothermal eel farms in Slovakia. Geothermal fisheries in Iceland grow arctic char, salmon, abalone, and other fish and shellfish. China has over 500 acres of geothermal fish farms, while Japanese fish farms grow eels and alligators. There are also fish farms in France, Greece, Israel, Korea, and New Zealand.

The main species raised in geothermal waters are catfish, bass, trout, tilapia, sturgeon, giant freshwater prawns, alligators, snails, coral, and tropical fish. The warmth of geothermal water makes it possible to raise tropical marine (saltwater) species in cold, land-locked places such as Idaho.

Some creatures have a range of temperatures in which they thrive. For example, catfish and shrimp grow at about 50 percent of optimum rate at temperatures between 68 and 79 °F (20 and 26 °C) and grow fastest at about 90 °F (32 °C), but they decline at temperatures higher than that. Trout thrive at around 60 °F (15.5 °C) but dislike lower or higher temperatures.

Scientists are investigating using geothermal aquaculture to grow plants that humans and animals could eat. Possible crops include kelp, duckweed, algae, and water hyacinth. As of the early 2000s, the technology was not yet good enough to allow economically worthwhile harvesting and processing.

Benefits and drawbacks of aquacultural applications

Like other direct uses of geothermal water, aquaculture allows an area to make use of groundwater that may not be hot enough to generate electricity but is still hot enough to be useful as hot water. Arizona, for example, has a great deal of geothermal water that is under 300 °F (149 °C), which cannot generate electricity but is very useful in aquaculture.

The fish grown in geothermal fisheries are healthier and stronger than fish grown in unheated fish ponds. Fish farmers can regulate temperature throughout the year to make sure the fish grow to a consistent size year-round.

However, fish farmers must be careful to regulate water temperature. The water in and near the pipes bringing in the hot groundwater can get very hot, creating pockets that are too hot for fish. For aquaculture to work well, there must be a source of cool water in addition to the hot water. Some geothermal fisheries collect geothermal water in holding ponds and let it cool in order to regulate pond temperatures. If the water does not circulate evenly there can also be cold spots. This can make the fish crowd into areas where the temperature is at the right level. The hot pipes also can be dangerous to human workers who must wade into the pools for repairs, feeding, and harvesting.

Impact of aquacultural applications

For the most part, using hot groundwater to heat fish ponds is good for the environment. A farm that uses geothermal water is not burning fossil fuels or other sources of heat to regulate water temperature and is therefore not emitting pollutants. Many geothermal aquaculture operations use water that has already been used by geothermal power plants or heating systems. The water has lost most of its heat but is still hot enough to raise the temperature of the fish ponds, so it can be put to a second use before disposal.

Aquaculture itself has both good and bad aspects for the environment. It takes pressure off wild fisheries, many of which have

been severely overfished. In some areas, however, it contributes to water pollution.

Economically, using geothermal energy to heat water for aquaculture can have many benefits. Places that use water that has already been used for heating or electricity generation can heat their fish ponds essentially for no cost. They can also enjoy the economic benefit of selling the fish or prawns that they produce. Fish grown in geothermally heated water grow faster than fish in unheated water, so some fish farmers can grow extra fish crops for sale. Heated water makes it possible to grow fish in winter when it ordinarily would not be possible. Selling tropical fish for the pet store market can be quite profitable. Developing nations can export their fish produce for good prices, bringing foreign capital into the country.

GEOTHERMAL POWER PLANTS

One very promising use of geothermal power is the generation of electricity. Areas with hot geothermal reservoirs can use this heat and steam to create electricity without having to spend money for fuel and without polluting the atmosphere or ground.

Geothermal power plant well in Dixie Valley near Fallon, Nevada. © *Inga Spence/ Visuals Unlimited. Reproduced by permission.*

All types of geothermal power plants use geothermal steam to turn a turbine. The turbine is attached to a generator that creates the electricity. The electricity is then fed into a grid, which is attached to individual users. There are three main types of geothermal power plant: binary, dry steam, and flashed steam. There are also hybrid power plants that combine geothermal energy with other energy sources. The type of plant built in a given area depends on what sort of geothermal resource is available, either steam or liquid and either high or low temperature.

Binary power plants

Binary power plants use a two-step process to extract power from geothermal water that is not quite hot enough to spin a turbine by itself. The hot water is pumped up through the ground and passed through a heat exchanger that contains a fluid with a much lower boiling point than water. The heat from the geothermal water causes this "binary" fluid to flash into vapor. That vapor spins the turbine, which powers the generator. The geothermal water is injected back into the reservoir. The binary fluid stays inside the tank, where it is used over and over again. Nothing is released into the atmosphere.

Many areas have geothermal reservoirs with water that is below 400 °F (204 °C). Moderate-temperature geothermal water is much more common than high-temperature water. The United States Department of Energy predicts that most geothermal power plants built in the future will be binary power plants that can take advantage of this slightly cooler water.

Dry steam plants

Dry steam plants use the steam that comes up from a geothermal reservoir to power turbines that power generators. The liquid and steam is then injected back into the reservoir to regain its heat and maintain the reservoir's pressure. Dry steam was the first technology used to build geothermal power plants. The plant built in Lardarello, Italy, in 1904 used dry steam technology. The Geysers in northern California uses dry steam to produce power. Dry steam is still the largest source of geothermal power in the world.

Flashed steam plants

Flashed steam plants are the most common type of geothermal power plant. These plants use geothermal water that is over 360 °F (182 °C). The fluid is pumped up at high pressure and then sprayed into a tank that is at lower pressure than the water. This causes the geothermal water to "flash," or turn into steam instantly. The steam

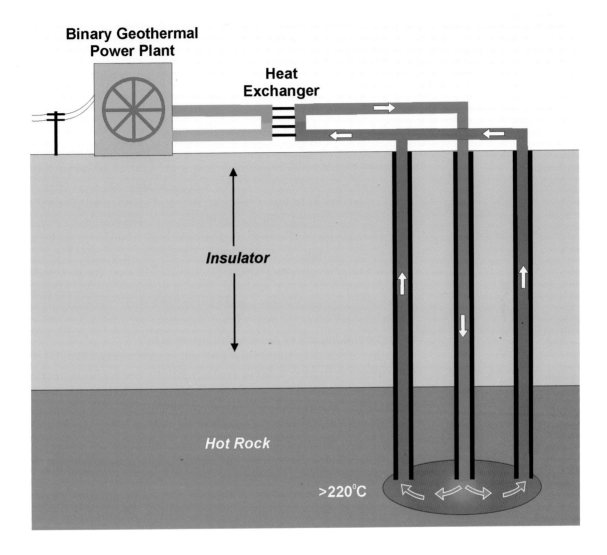

**Binary Geothermal
Power Plant**

**Heat
Exchanger**

Insulator

Hot Rock

>220°C

spins a turbine, which powers a generator. Fluid left in the first tank is then pumped into another tank to be flashed again. After the water has been used, it is injected back into the reservoir to regain its heat.

Hybrid power plants

Some areas do not have enough geothermal energy to run a full power plant. These places can be ideal sites for hybrid power plants that combine different types of power generation. They can combine different types of geothermal energy generation or combine geothermal energy with other energy sources, even fossil fuels.

A graphic illustration shows a geothermal technology Australian companies are developing to generate electricity from the heat of ancient rocks buried deep below the red sands of the Australian outback. Spurred by high commodity prices and a drive to reduce Australia's reliance on coal, several

Benefits and drawbacks of geothermal power plants

Geothermal power plants are usually built with modular designs, which makes them very flexible. It is easy to start with a small plant and then add additional units if the demand for electricity increases. Geothermal plants can also use some of their water, either freshly pumped or after being used for electricity, for other direct purposes, such as heating or aquaculture.

However, not every location can use geothermal power. Geothermal power plants must be located near a geothermal reservoir that has water of at least 250 °F (121 °C) and preferably 300 °F (148 °C). Not all reservoirs have water this hot. An ideal geothermal reservoir is hot with low mineral content, has shallow aquifers nearby to make it easy to re-inject used water, is on private land in order to make it easier to get permits, is near existing electrical transmission lines, and has a supply of cooler water for cooling. It also needs a high enough volume of water to keep flowing steadily. In the United States, only the western states and Hawaii have these resources.

Environmental impact of geothermal power plants

Geothermal power plants are generally environmentally clean. They do not burn fossil fuels, so they help conserve those fuels for other purposes. They produce no emissions to contribute to air pollution, the greenhouse effect, or global warming. There is no smoke surrounding geothermal power plants. Dry steam and flashed steam plants emit excess steam and small amounts of gases, while binary plants emit nothing at all because all the fluids are contained within the system and recycled. Areas that use geothermal power have some of the best air quality readings in the world. Lake County, California, which has five geothermal power plants, is the only county in the United States that has met the strictest governmental air quality standards since the mid-1990s.

Geothermal plants do not need space to store fuels, and they do not create large piles of ash that must be cleared or oil spills that damage oceans. They also do not pollute groundwater, unless the geothermal water has a high concentration of minerals or metals.

Unlike most other power plant types, geothermal plants do not require large amounts of space to function. They can be built right on top of geothermal reservoirs. The pumps that bring water up from geothermal reservoirs are small, especially compared to those used by coal mines or oil wells. They do not tear up large plots of land or destroy forests. There is no need to build major highways, railroads, or pipelines in order to transport fuel to geothermal power plants because their source of power is directly below them. It is actually possible to build geothermal power plants in the midst of farmland or forests, where they can coexist with livestock and wildlife.

Economic impact

Geothermal power plants require an initial investment in finding reservoirs, digging wells, and building a plant with turbines. This initial investment can be quite heavy, between $3,000 and $5,000 per kilowatt. Once a plant is built, however, it can be more economical than producing power with fossil fuels. Fuel does not have to be purchased to run the plant, which saves money and makes operations more predictable, as the plant is not affected by fluctuations in the price of oil, gas, or coal.

Issues, challenges, and obstacles of geothermal power plants

In developed nations the existing utility companies have a large investment in their currently functioning power plants. These plants usually run on fossil fuels, though there are some nuclear and hydroelectric power plants. The utilities themselves and the oil and gas companies that supply their fuel have an interest in maintaining things as they are. There is little incentive for them to give up their source of income in favor of geothermal power.

Geothermal power plants can only be built on or near geothermal reservoirs. These reservoirs are often on private land or land that is already being used for some other purpose. A company that wants to build a geothermal power plant must first get access to the land over the reservoir, which can be difficult, expensive, and time-consuming. There needs to be a great deal more research and development before geothermal power generation becomes practical around the world.

GEOTHERMAL HEATING APPLICATIONS

One obvious use of geothermal energy is for heat. Many cities and homes use naturally hot water to keep them warm in winter. There are two main ways to use geothermal water for heating. The older method is using the water directly. Newer technology involves using a geothermal heat pump.

Direct heating

Direct heating pumps the water from the geothermal reservoir in the ground and passes it through pipes running through buildings.

Oregon's Geothermal Zone

Klamath Falls, Oregon, has used geothermal heating for homes since 1900. In the early 2000s, more than 550 geothermal wells were in use, heating homes, pools, schools, and businesses. Geothermal pipes run under the sidewalks and highways to keep them clear of snow. In 1982 the city built a geothermal district heating system that heats the entire eastern part of the city. Two wells east of downtown pump water that is about 210 °F (98 °C) from underground reservoirs to the central mechanical room at the County Museum. This water is treated and then delivered to customers. It is about 180 °F (82 °C) when it reaches the seven hundred homes and buildings that use geothermal heat. When it returns to the mechanical room, it has lost about 40 °F (4 °C) of temperature. It is then injected back into the reservoir to be recycled.

The heat from the water moves from the pipes through the walls into the air inside the building. This system can also be used to heat water.

For direct heating, the best geothermal water temperature is under 212 °F (100 °C). In fact, water with a temperature as low as 95 °F (35 °C) can be used for direct heating. In some areas, such as Iceland, the geothermal water is pure enough that it can be pumped directly through radiators. In most places, however, chemicals in the water make it necessary to filter the water through heat exchangers that extract the heat from the water.

Geothermal heat pumps

Newer technology uses geothermal water to run a heat pump, similar to an electric heat pump. A geothermal heat pump forces heat in a direction it would not ordinarily go. Most heat pumps can function as both heating and cooling units. In winter they heat air and pump it through the house. In summer they absorb hot air and pump it into the ground. Geothermal heat pumps are particularly efficient because they start with air or water that is already hot and thus do not have to heat it as much as ordinary heat pumps, which start with cold outside air. Geothermal heat pumps use 30 to 60 percent less electricity than traditional heat pumps because they do not have to create their own heat, just move it from place to place.

Geothermal heat pumps work by pumping water or a mix of water and antifreeze through the ground next to a house or building. The ground temperature remains relatively constant throughout the year, generally between 45 and 55 °F (7 and 12 °C). In winter, underground pipes absorb heat from the Earth. This heated water circulates into the heat pump, where it is concentrated so that it will increase to the desired room temperature. The heat pump then pumps the hot air through the ducts in the building, heating the rooms. In summer the process is reversed; the hot air is sucked from the building and dispersed into the ground. The geothermal heat pump system uses ordinary ductwork, so there is no need to modify existing ducts.

Geothermal heat pumps usually can at least partially heat water for the home. This is not necessarily possible all year round. During

Reykjavik is the capital of Iceland. Nearly all of the hot water in Reykjavik is obtained from natural geothermal sources. Much of its energy is derived from natural sources such as geothermal and hydroelectric power. These energy sources are non-polluting and essentially inexhaustible, making Iceland a clean and environmentally friendly country. *Martin Bond/Photo Researchers, Inc.*

the summer the heat pump can use excess heat to warm domestic hot water, but during the winter there is not as much heat available to warm the water. A home with a geothermal heat pump must usually have an alternate source of heat for water, but even so, using excess heat for even part of the year results in an energy savings. New technology is improving this situation; because geothermal heat pumps are so much more efficient than other forms of water heating, some manufacturers are now selling geothermal heat pumps that heat water separately, thereby providing hot water year-round.

Geothermal heat pumps, like all heat pumps, produce slightly warmer air than fossil fuel furnaces. Geothermal heat pumps generally produce hot air between 95 and 103 °F (35 and 39 °C), as opposed to conventional heat pumps, which produce hot air between 90 and 95 °F (32 and 35 °C). Geothermal heat pumps require more open ductwork for air flow than fossil fuel furnaces, which can be a problem when converting older houses to geothermal heat.

Current uses of geothermal heating applications

People have used geothermal water to heat buildings for hundreds of years. People in Paris, France, heated buildings with geothermal water six hundred years ago. Boise, Idaho, began using geothermal heating in 1892. This system is still in use there, where four district heating systems heat over five million square feet (152 million square meters) of space.

Starting in the 1960s, other cities began to take notice of the potential benefits of geothermal energy. By the early 2000s geothermal direct heating was common in Iceland, Hungary, Poland, China, Argentina, Croatia, France, and Turkey. Reykjavik, Iceland, has the world's largest geothermal heating system, with about two hundred miles (320 kilometers) of pipes running throughout the city. The city is almost entirely heated by geothermal heat.

Geothermal heat pumps are gradually gaining popularity as people learn about them. Geothermal heat pumps in the early 2000s were considered much more efficient than the ones made in 1990. Experts foresee some continuing improvements but believe they will be small compared to improvements already made.

Benefits and drawbacks of geothermal heating applications

Geothermal direct heat is inexpensive and nonpolluting. Places that have sufficient geothermal resources can heat entire cities for just the cost of running the pipes. The heat is always available and does not depend on fuel supplies. However, geothermal direct heat is only possible in areas with substantial geothermal resources. That means it cannot be a worldwide solution to the heating problem.

On the other hand, geothermal heat pumps can be used almost anywhere in the world because they do not require the presence of geothermal reservoirs. They make it possible to use geothermal resources that were formerly considered unusable. They can be used for summer cooling in addition to winter heating, and sometimes they can supply hot water as well. These pumps are easiest to install in new buildings; it is difficult to convert existing homes to geothermal heat pumps, and they cost more than electric heat pumps.

Impact of geothermal heating applications

Geothermal heating has many obvious environmental benefits. It does not pollute the air at all because geothermal heating involves no combustion and therefore no emissions. Geothermal heat pumps pose few environmental problems. They use an anti-freeze substance, but it is usually a nontoxic chemical called propylene glycol or small amounts of methanol, both of which are commonly used in windshield washing solutions.

Economically, geothermal heat can be much less expensive than other sources of heat, such as fossil fuels or wood, but that cost depends on several factors. The initial installation costs can be high, but if the heating system works well it can pay for itself quickly. Geothermal heat works especially well in areas that already have wells dug into geothermal reservoirs. If there are already wells in place, a district or institution only needs to buy pipelines, heat exchangers, and pumps. A heating system is more expensive to install if there is not already a good geothermal reservoir in use.

Geothermal heat pumps currently cost more than conventional ones, but once they are installed the cost of running them is less than that of any other conventional form of heat, including natural gas. This savings depends on the cost of fossil fuels; as fossil fuels get more expensive, geothermal heat may become more economical. It is estimated that geothermal heat pumps can reduce the power used to heat or cool a house by one to five kilowatts of generating capacity at peak time, which can result in major savings on residential heating and cooling costs.

Issues, challenges, and obstacles of geothermal heating applications

Experts estimate that almost three hundred communities in the western United States are close enough to geothermal heat sources to use them as a district heating system. Many other countries have the potential to use more geothermal energy for district heating. People are gradually taking more interest in geothermal resources as fossil fuels become more expensive and the dangers of air pollution become more apparent.

Implementing a geothermal heating system is a major investment. It requires money, labor, and a willingness to take the risk that it may not work. The technology is still new and does not have a long track record, nor are there many people who are experts in installing geothermal heating systems. There have been unsuccessful attempts to use geothermal heat.

In the early 2000s, there were about 500,000 geothermal heat pumps in use in the United States. Switzerland and several other countries were implementing programs to increase geothermal heat pump usage. There is plenty of potential for expansion. People do not use them mainly because they are not widely available, and few people know that they exist. There is also the problem of persuading people to buy geothermal heat pumps when they cost more than conventional climate control systems.

INDUSTRIAL APPLICATIONS

Many industries need steam or hot water for their operations. Geothermal water is an excellent low-cost source of this basic item. Industries generally need very hot water, hotter than the water used in agriculture or aquaculture, though there is much variation. Plants can be built right next to geothermal reservoirs and pipe the water or steam straight into the operation.

Current uses of industrial applications

Geothermal water is useful in any industry that requires steam or hot water. Some uses include:

- Timber processing
- Pulp and paper processing
- Washing wool
- Dyeing cloth
- Drying diatomaceous earth (a light, abrasive soil used as a filtering material and insecticide)
- Drying fish meal and stock fish
- Canning food
- Drying cement

Geothermal Dye Works

In most cases an industry uses geothermal water because it is a cheap source of heat and/or water. In a few cases, however, industries take advantage of the unique mineral properties of geothermal water. In Iwate Prefecture, Japan, there is a new geothermal dye factory that uses the minerals in geothermal water as a mordant, a substance that makes dye pigments stick to cloth, and also as a substance that can remove dye from cloth. The factory uses a method of folding and tying the cloth with string, soaking it in dye made with geothermal water, and then rinsing it and unfolding it. The combination of steam, heat, and the hydrogen sulfide in the geothermal water leaves beautiful and unique patterns on the cloth.

- Drying organic materials such as vegetables, seaweed, and grass
- Refrigeration

Benefits and drawbacks of industrial applications

Using geothermal water and steam saves companies the cost of heating water and saves the environment some of the pollution that would be caused by heating the water. However, geothermal water is only available in a few places, so most industries cannot use it.

Issues, challenges, and obstacles

The use of geothermal water in industry is still very new, and as of the early 2000s, not many industries are taking advantage of it. Few people know whether or not geothermal energy is available or how to use it if it is. Implementing geothermal energy requires installing equipment such as pipes, which can be expensive or difficult in an existing plant.

For More Information

Books

Cataldi, Raffaele, ed. *Stories from a Heated Earth: Our Geothermal Heritage.* Davis, CA: Geothermal Resources Council, 1999.

Dickson, Mary H., and Mario Fanelli, eds. *Geothermal Energy: Utilization and Technology*. London: Earthscan Publications, 2005.

Geothermal Development in the Pacific Rim. Davis, CA: Geothermal Resources Council, 1996.

Graham, Ian. *Geothermal and Bio-Energy*. Fort Bragg, CA: Raintree, 1999.

Wohltez, Kenneth, and Grant Keiken. *Volcanology and Geothermal Energy*. Berkeley: University of California Press, 1992.

Periodicals
Anderson, Heidi. ''Environmental Drawbacks of Renewable Energy: Are They Real or Exaggerated?'' *Environmental Science and Engineering* (January 2001).

Web sites
''Geo-Heat Center.'' Oregon Institute of Technology. http://geoheat.oit.edu. (accessed on July 19, 2005).

''Geothermal Energy.'' World Bank. http://www.worldbank.org/html/fpd/energy/geothermal. (accessed on July 19, 2005).

Geothermal Energy Association. http://www.geo-energy.org. (accessed on August 4, 2005).

Geothermal Resources Council. http://www.geothermal.org. (accessed on August 4, 2005).

''Geothermal Technologies Program.'' U.S. Department of Energy: Energy Efficiency and Renewable Energy. http://www.eere.energy.gov/geothermal. (accessed on July 22, 2005).

Other sources
World Spaceflight News. *21st Century Complete Guide to Geothermal Energy* (CD-ROM). Progressive Management, 2004.

Where to Learn More

BOOKS

Angelo, Joseph A. *Nuclear Technology.* Westport, CT: Greenwood Press, 2004.

Avery, William H., and Chih Wu. *Renewable Energy from the Ocean.* New York: Oxford University Press, 1994.

Berinstein, Paula. *Alternative Energy: Facts, Statistics, and Issues.* Phoenix, AZ: Oryx Press, 2001.

Boyle, Godfrey. *Renewable Energy,* 2nd ed. New York: Oxford University Press, 2004.

Buckley, Shawn. *Sun Up to Sun Down: Understanding Solar Energy.* New York: McGraw-Hill, 1979.

Burton, Tony, David Sharpe, Nick Jenkins, and Ervin Bossanyi. *Wind Energy Handbook.* New York: Wiley, 2001.

Carter, Dan M., and Jon Halle. *How to Make Biodiesel.* Winslow, Bucks, UK: Low-Impact Living Initiative (Lili), 2005.

Cataldi, Raffaele, ed. *Stories from a Heated Earth: Our Geothermal Heritage.* Davis, CA: Geothermal Resources Council, 1999.

Close, Frank E. *Too Hot to Handle: The Race for Cold Fusion.* Princeton, NJ: Princeton University Press, 1991.

Cook, Nick. *The Hunt for Zero Point: Inside the Classified World of Antigravity Technology.* New York: Broadway Books, 2003.

Cuff, David J., and William J. Young. *The United States Energy Atlas,* 2nd ed. New York: Macmillan, 1986.

Dickson, Mary H., and Mario Fanelli, eds. *Geothermal Energy: Utilization and Technology.* London: Earthscan Publications, 2005.

Domenici, Peter V. *A Brighter Tomorrow : Fulfilling the Promise of Nuclear Energy.* Lanham, MD: Rowman and Littlefield, 2004.

Ewing, Rex. *Hydrogen: Hot Cool Science—Journey to a World of the Hydrogen Energy and Fuel Cells at the Wassterstoff Farm.* Masonville, CO: Pixyjack Press, 2004.

Freese, Barbara. *Coal: A Human History.* New York: Perseus, 2003.

Frej, Anne B. *Green Office Buildings: A Practical Guide to Development.* Washington, DC: Urban Land Institute, 2005.

Gelbspan, Ross. *Boiling Point: How Politicians, Big Oil and Coal, Journalists and Activists Are Fueling the Climate Crisis.* New York: Basic Books, 2004.

Geothermal Development in the Pacific Rim. Davis, CA: Geothermal Resources Council, 1996.

Graham, Ian. *Geothermal and Bio-Energy.* Fort Bragg, CA: Raintree, 1999.

Heaberlin, Scott W. *A Case for Nuclear-Generated Electricity: (Or Why I Think Nuclear Power Is Cool and Why It Is Important That You Think So Too).* Columbus, OH: Battelle Press, 2003.

Howes, Ruth, and Anthony Fainberg. *The Energy Sourcebook: A Guide to Technology, Resources and Policy.* College Park, MD: American Institute of Physics, 1991.

Husain, Iqbal. *Electric and Hybrid Vehicles: Design Fundamentals.* Boca Raton, FL: CRC Press, 2003.

Hyde, Richard. *Climate Responsive Design.* London: Taylor and Francis, 2000.

Kaku, Michio, and Jennifer Trainer, eds. *Nuclear Power: Both Sides.* New York: Norton, 1983.

Kibert, Charles J. *Sustainable Construction: Green Building Design and Delivery.* New York: Wiley, 2005.

Leffler, William L. *Petroleum Refining in Nontechnical Language.* Tulsa, OK: Pennwell Books, 2000.

Lusted, Marcia, and Greg Lusted. *A Nuclear Power Plant.* San Diego, CA: Lucent Books, 2004.

Manwell, J. F., J. G. McGowan, and A. L. Rogers. *Wind Energy Explained.* New York: Wiley, 2002.

McDaniels, David K. *The Sun.* 2nd ed. New York: John Wiley & Sons, 1984.

Morris, Robert C. *The Environmental Case for Nuclear Power.* St. Paul, MN: Paragon House, 2000.

National Renewable Energy Laboratory, U.S. Department of Energy. *Wind Energy Information Guide.* Honolulu, HI: University Press of the Pacific, 2005.

Ord-Hume, Arthur W. J. G. *Perpetual Motion: The History of an Obsession.* New York: St. Martin's Press, 1980.

Pahl, Greg. *Biodiesel: Growing a New Energy Economy.* Brattleboro, VT: Chelsea Green Publishing Company, 2005.

Rifkin, Jeremy. *The Hydrogen Economy.* New York: Tarcher/Putnam, 2002.

Romm, Joseph J. *The Hype of Hydrogen: Fact and Fiction in the Race to Save the Climate.* Washington, DC: Island Press, 2004.

Seaborg, Glenn T. *Peaceful Uses of Nuclear Energy.* Honolulu, HI: University Press of the Pacific, 2005.

Tickell, Joshua. *From the Fryer to the Fuel Tank: The Complete Guide to Using Vegetable Oil as an Alternative Fuel.* Covington, LA: Tickell Energy Consultants, 2000.

Wohltez, Kenneth, and Grant Keiken. *Volcanology and Geothermal Energy.* Berkeley: University of California Press, 1992.

Wulfinghoff, Donald R. *Energy Efficiency Manual: For Everyone Who Uses Energy, Pays for Utilities, Designs and Builds, Is Interested in Energy Conservation and the Environment.* Wheaton, MD: Energy Institute Press, 2000.

PERIODICALS

Anderson, Heidi. "Environmental Drawbacks of Renewable Energy: Are They Real or Exaggerated?" *Environmental Science and Engineering* (January 2001).

Behar, Michael. "Warning: The Hydrogen Economy May Be More Distant Than It Appears." *Popular Mechanics* (January 1, 2005): 64.

Brown, Kathryn. "Invisible Energy." *Discover* (October 1999): 36.

Burns, Lawrence C., J. Byron McCormick, and Christopher E. Borroni-Bird. "Vehicles of Change." *Scientific American* (October 2002): 64-73.

Corcoran, Elizabeth. "Bright Ideas." *Forbes* (November 24, 2003): 222.

Dixon, Chris. "Shortages Stifle a Boom Time for the Solar Industry." *New York Times* (August 5, 2005): A1.

Feldman, William. "Lighting the Way: To Increased Energy .Efficiency." *Journal of Property Management* (May 1, 2001): 70.

Freeman, Kris. "Tidal Turbines: Wave of the Future?" *Environmental Health Sciences* (January 1, 2004): 26.

Graber, Cynthia. "Building the Hydrogen Boom." *OnEarth* (Spring 2005): 6.

Grant, Paul. "Hydrogen Lifts Off—with a Heavy Load." *Nature* (July 10, 2003): 129-130.

Guteral, Fred, and Andrew Romano. "Power People." *Newsweek* (September 20, 2004): 32.

Hakim, Danny. "George Jetson, Meet the Sequel." *New York Times* (January 9, 2005): section 3, p. 1.

Lemley, Brad. "Lovin' Hydrogen." *Discover* (November 2001): 53-57, 86.

Libby, Brian. "Beyond the Bulbs: In Praise of Natural Light." *New York Times* (June 17, 2003): F5.

Linde, Paul. "Windmills: From Jiddah to Yorkshire." *Saudi Aramco World* (January/February 1980). This article can also be found online at http://www.saudiaramcoworld.com/issue/198001/windmills-from.jiddah.to.yorkshire.htm.

Lizza, Ryan. "The Nation: The Hydrogen Economy; A Green Car That the Energy Industry Loves." *New York Times* (February 2, 2003): section 4, p. 3.

McAlister, Roy. "Tapping Energy from Solar Hydrogen." *World and I* (February 1999): 164.

Motavalli, Jim. "Watt's the Story? Energy-Efficient Lighting Comes of Age." *E* (September 1, 2003): 54.

Muller, Joann, and Jonathan Fahey. "Hydrogen Man." *Forbes* (December 27, 2004): 46.

Nowak, Rachel. "Power Tower." *New Scientist* (July 31, 2004): 42.

Parfit, Michael. "Future Power: Where Will the World Get Its Next Energy Fix?" *National Geographic* (August 2005): 2–31.

Pearce, Fred. "Power of the Midday Sun." *New Scientist* (April 10, 2004): 26.

Perlin, John. "Soaring with the Sun." *World and I* (August 1999): 166.

Port, Otis. "Hydrogen Cars Are Almost Here, but . . . There Are Still Serious Problems to Solve, Such As: Where Will Drivers Fuel Up?" *Business Week* (January 24, 2005): 56.

Provey, Joe. "The Sun Also Rises." *Popular Mechanics* (September 2002): 92.

Service, Robert F. "The Hydrogen Backlash." *Science* (August 13, 2004): 958-961.

"Stirrings in the Corn Fields." *The Economist* (May 12, 2005).

Terrell, Kenneth. "Running on Fumes." *U.S. News & World Report* (April 29, 2002): 58.

Tompkins, Joshua. "Dishing Out Real Power." *Popular Science* (February 1, 2005): 31.

Valenti, Michael. "Storing Hydroelectricity to Meet Peak-Hour Demand." *Mechanical Engineering* (April 1, 1992): 46.

Wald, Matthew L. "Questions about a Hydrogen Economy." *Scientific American* (May 2004): 66.

Westrup, Hugh. "Cool Fuel: Will Hydrogen Cure the Country's Addiction to Fossil Fuels?" *Current Science* (November 7, 2003): 10.

Westrup, Hugh. "What a Gas!" *Current Science* (April 6, 2001): 10.

WEB SITES

"Alternative Fuels." U.S. Department of Energy Alternative Fuels Data Center. http://www.eere.energy.gov/afdc/altfuel/altfuels.html (accessed on July 20, 2005).

"Alternative Fuels Data Center." U.S. Department of Energy: Energy Efficiency and Renewable Energy. http://www.eere.energy.gov/afdc/altfuel/p-series.html (accessed on July 11, 2005).

American Council for an Energy-Efficient Economy. http://aceee.org/ (accessed on July 27, 2005).

The American Solar Energy Society. http://www.ases.org/ (accessed on September 1, 2005).

Biodiesel Community. http://www.biodieselcommunity.org/ (accessed on July 27, 2005).

"Bioenergy." Natural Resources Canada. http://www.canren.gc.ca/tech_appl/index.asp?CaId=2&PgId=62 (accessed on July 29, 2005).

"Biofuels." Journey to Forever. http://journeytoforever.org/biofuel.html (accessed on July 13, 2005).

"Biogas Study." Schatz Energy Research Center. http://www.humboldt.edu/~serc/biogas.html (accessed on July 15, 2005).

"Black Lung." United Mine Workers of America. http://www.umwa.org/blacklung/blacklung.shtml (accessed on July 20, 2005).

"Classroom Energy!" American Petroleum Institute. http://www.classroom-energy.org (accessed on July 20, 2005).

"Clean Energy Basics: About Solar Energy." National Renewable Energy Laboratory. http://www.nrel.gov/clean_energy/solar.html (accessed on August 25, 2005).

"A Complete Guide to Composting." Compost Guide. http://www.compostguide.com/ (accessed on July 25, 2005).

"Conserval Engineering, Inc." American Institute of Architects. http://www.solarwall.com/ (accessed on September 1, 2005).

"The Discovery of Fission." Center for History of Physics. http://www.aip.org/history/mod/fission/fission1/01.html (accessed on December 17, 2005).

"Driving for the Future." California Fuel Cell Partnership. http://www.cafcp.org (accessed on August 8, 2005).

"Driving and Maintaining Your Vehicle." Natural Resources Canada. http://oee.nrcan.gc.ca/transportation/personal/driving/autosmart-maintenance.cfm?attr=11 (accessed on September 28, 2005).

"Ecological Footprint Quiz." Earth Day Network. http://www.earthday.net/footprint/index.asp (accessed on February 6, 2006).

"Energy Efficiency and Renewable Energy." U.S. Department of Energy. http://www.eere.energy.gov (accessed on September 28, 2005).

"Ethanol: Fuel for Clean Air." Minnesota Department of Agriculture. http://www.mda.state.mn.us/ethanol/ (accessed on July 14, 2005).

"Florida Solar Energy Center." University of Central Florida. http://www.fsec.ucf.edu (accessed on September 1, 2005).

"Fueleconomy.gov." United States Department of Energy. http://www.fueleconomy.gov/feg/ (accessed on July 27, 2005).

Fukada, Takahiro. "Japan Plans To Launch Solar Power Station In Space By 2040." *SpaceDaily.com,* Jan. 1, 2001. Available at http://www.spacedaily.com/news/ssp-01a.html (accessed Feb. 12, 2006).

"Geo-Heat Center." Oregon Institute of Technology. http://geoheat.oit.edu. (accessed on July 19, 2005).

"Geothermal Energy." World Bank. http://www.worldbank.org/html/fpd/energy/geothermal. (accessed on July 19, 2005).

Geothermal Resources Council. http://www.geothermal.org. (accessed on August 4, 2005).

"Geothermal Technologies Program." U.S. Department of Energy: Energy Efficiency and Renewable Energy. http://www.eere.energy.gov/geothermal. (accessed on July 22, 2005).

"Green Building Basics." California Home. http://www.ciwmb.ca.gov/GreenBuilding/Basics.htm (accessed on September 28, 2005).

"Guided Tour on Wind Energy." Danish Wind Industry Association. http://www.windpower.org/en/tour.htm (accessed on July 25, 2005).

"How the BMW H2R Works." How Stuff Works. http://auto.howstuffworks.com/bmw-h2r.htm (accessed on August 8, 2005).

"Hydrogen, Fuel Cells & Infrastructure Technologies Program." U.S. Department of Energy Energy Efficiency and Renewable Energy. http://www.eere.energy.gov/hydrogenandfuelcells/ (accessed on August 8, 2005).

"Hydrogen Internal Combustion." Ford Motor Company. http://www.ford.com/en/innovation/engineFuelTechnology/hydrogenInternalCombustion.htm (accessed on August 8, 2005).

"Incandescent, Fluorescent, Halogen, and Compact Fluorescent." California Energy Commission. http://www.consumerenergy-center.org/homeandwork/homes/inside/lighting/bulbs.html (accessed on September 28, 2005).

"Introduction to Green Building." Green Roundtable. http://www.greenroundtable.org/pdfs/Intro-To-Green-Building.pdf (accessed on September 28, 2005).

Lovins, Amory. "Mighty Mice: The most powerful force resisting new nuclear may be a legion of small, fast and simple microgeneration and efficiency projects." *Nuclear Engineering International*, Dec. 2005. Available at http://www.rmi.org/images/other/Energy/E05-15_MightyMice.pdf (accessed Feb. 12, 2006).

Nice, Karim. "How Hybrid Cars Work." Howstuffworks.com. http://auto.howstuffworks.com/hybrid-car.htm (accessed on September 28, 2005).

"Nuclear Terrorism—How to Prevent It." Nuclear Control Institute. http://www.nci.org/nuketerror.htm (accessed on December 17, 2005).

"Oil Spill Facts: Questions and Answers." *Exxon Valdez* Oil Spill Trustee Council. http://www.evostc.state.ak.us/facts/qanda.html (accessed on July 20, 2005).

O'Mara, Katrina, and Mark Rayner. "Tidal Power Systems." http://reslab.com.au/resfiles/tidal/text.html (accessed on September 13, 2005).

"Photos of El Paso Solar Pond." University of Texas at El Paso. http://www.solarpond.utep.edu/page1.htm (accessed on August 25, 2005).

"The Plain English Guide to the Clean Air Act." U.S. Environmental Protection Agency. http://www.epa.gov/air/oaqps/peg_caa/pegcaain.html (accessed on July 20, 2005).

"Reinventing the Automobile with Fuel Cell Technology." General Motors Company. http://www.gm.com/company/gmability/adv_tech/400_fcv/ (accessed on August 8, 2005).

"Safety of Nuclear Power." Uranium Information Centre, Ltd. http://www.uic.com.au/nip14.htm (accessed on December 17, 2005).

"Solar Energy for Your Home." Solar Energy Society of Canada Inc. http://www.solarenergysociety.ca/2003/home.asp (accessed on August 25, 2005).

The Solar Guide. http://www.thesolarguide.com (accessed on September 1, 2005).

"Solar Ponds for Trapping Solar Energy." United National Environmental Programme. http://edugreen.teri.res.in/explore/renew/pond.htm (accessed on August 25, 2005).

"Thermal Mass and R-value: Making Sense of a Confusing Issue." BuildingGreen.com. http://buildggreen.com/auth/article.cfm?fileName=070401a.xml (accessed on September 28, 2005).

"Tidal Power." University of Strathclyde. http://www.esru.strath.ac.uk/EandE/Web_sites/01-02/RE_info/Tidal%20Power.htm (accessed on September 13, 2005).

U.S. Department of Energy. *Report of the Review of Low Energy Nuclear Reactions.* Washington, DC: Department of Energy, Dec. 1, 2004. http://lenr-canr.org/acrobat/DOEreportofth.pdf, accessed Feb. 12, 2006.

Vega, L. A. "Ocean Thermal Energy Conversion (OTEC)." http://www.hawaii.gov/dbedt/ert/otec/index.html (accessed on September 13, 2005).

Venetoulis, Jason, Dahlia Chazan, and Christopher Gaudet. "Ecological Footprint of Nations: 2004." Redefining Progress. http://www.rprogress.org/newpubs/2004/footprintnations2004.pdf (accessed on February 8, 2006.)

Weiss, Peter. "Oceans of Electricity." *Science News Online* (April 14, 2001). http://www.science news.org/articles/20010414/bob12.asp (accessed on September 13, 2005).

"What Is Uranium? How Does It Work?" World Nuclear Association. http://www.world-nuclear.org/education/uran.htm (accessed on December 17, 2005).

"Wind Energy Tutorial." American Wind Energy Association. http://www.awea.org/faq/index.html (accessed on July 25, 2005).

Yam, Philip. "Exploiting Zero-Point Energy." *Scientific American,* December 1997. Available from http://www.padrak.com/ine/ZPESCIAM.html (accessed on August 2, 2005).

OTHER SOURCES

World Spaceflight News. *21st Century Complete Guide to Hydrogen Power Energy and Fuel Cell Cars: FreedomCAR Plans, Automotive Technology for Hydrogen Fuel Cells, Hydrogen Production, Storage, Safety Standards, Energy Department, DOD, and NASA Research.* Progressive Management, 2003.

■■■

Index

Italic type indicates volume number; **boldface type indicates entries and their pages;** *(ill.) indicates illustrations.*

A

Abraham, Spencer, *2:* 143
AC (Alternating current), *2:* 243
Acetone, *3:* 406
Acetylene, *1:* 48
Acid rain, *1:* 14, 15-16, 41
Active solar systems, *2:* 218
Adams, William G., *2:* 212, 214
Adobe, *2:* 210, *3:* 349
Aerodynamics, *3:* 368-69
Aeromotor Company, *3:* 311-12
Afghanistan
   hydropower in, *3:* 277
   windmills in, *3:* 306
Agriculture
   for biofuels, *1:* 66-67
   geothermal energy for, *1:* 115-18
   hydroelectric dams and, *3:* 286
   open field, *1:* 115-16
   See also Farms
Air conditioning
   natural gas for, *1:* 35
   ocean thermal energy conversion for, *3:* 292
   See also Cooling
Air pollution, *1:* 17 (ill.)
   from biofuels, *1:* 65

from coal, *1:* 20, 40-42, 43, *2:* 202, *3:* 340
from coal gasification, *1:* 45
from ethanol, *1:* 91
from exhaust emissions, *1:* 6, 12, 93, *2:* 142-43
from fossil fuels, *1:* 12-15, 20
from gasoline, *1:* 28
indoor, *1:* 14-15, 16, *3:* 347, 348
from methane, *1:* 86-87
from methanol, *1:* 51-52
MTBE and, *1:* 53
from natural gas, *1:* 37
nuclear power plants and, *2:* 202
particulate matter, *1:* 12-13, 14, 16
from P-Series fuels, *1:* 93
sources of, *1:* 14
Air quality standards, *1:* 124
Airplanes
   history of, *1:* 10
   lift and drag on, *3:* 324-25
   solar-powered, *2:* 219
   weight on, *3:* 373
   zero point energy for, *3:* 395
Airships, *2:* 137-39, 140 (ill.), 141, 142 (ill.)
Akron (Airship), *2:* 138
Alberta, Canada oil sands, *1:* 25

Alcohol fuels, *1:***87-92**
Al-Dimashqi, *3:* 308
Alexander the Great, *1:* 41
Algae
   biodiesel from, *1:* 75
   hydrogen from, *2:* 149
Alkaline fuel cells, *2:* 145-46
al-Qaeda, *2:* 199
Alternating current (AC), *2:* 243
Alternative energy, *3:* 380-84
   See also Renewable energy;
      specific types of alternative
      energy
Aluminum cans, recycled, *3:* 376
American Society of Civil Engineers, *3:* 283
American Wind Energy Association, *3:* 319
Ammonia
   biogas, *1:* 84
   for ocean thermal energy conversion, *3:* 290, 293
   in solar collectors, *2:* 215
Amorphous photovoltaic cells, *2:* 237-38
Anaerobic digestion technology, *1:* 84
Anasazi Indians, *3:* 343
Anemometers, *3:* 327
Animal waste. *See* Dung; Manure